Terminate the Controversy over the Big Bang Theory by Inspecting All Its Three Pillars

ALSO BY BINGCHENG ZHAO

(Popular Science & Science)

Why It's Difficult to Understand "A Brief History of Time"

The Long Night Is Over: The Mystery of Dark Matter Has Been Unveiled; The Constituents of Dark Matter Have Been Revealed

From Postulate-Based Modern Physics to Mechanism-Revealed Physics

(Science & Religion)

The Amazing Wisdom and Truth: GOD Can Change the Scales of Space and Time in the Universe

The Naked Truth: The Big Bang Theory Cannot Deny GOD!

Terminate the Controversy over the Big Bang Theory by Inspecting All Its Three Pillars

Bingcheng Zhao

TERMINATE THE CONTROVERSY OVER THE BIG
BANG THEORY BY INSPECTING ALL ITS THREE PILLARS

Published in the United States of America through Amazon CreateSpace

The copyright of this book is owned by the author of this book

Copyright © 2018 by Bingcheng Zhao—the author of this book

All rights reserved, including the right of reproduction in whole or in part in any form or by any means. No part of this book may be reproduced or transmitted in any form or by any means—electronic or mechanical, photocopying, recording, without the permission from the author of this book. No part of this book may be translated into any other languages without the permission from the author of this book.

ISBN-13: 978-1727281880
ISBN-10: 1727281888

CONTENTS

FOREWORD ... vii

Chapter 1: Inspect the Crucial Pillar of the Big Bang Theory ... 1
    The prerequisite for this crucial pillar to be valid ... 1
    Inspect whether this prerequisite is satisfied from four angles ... 4
    The inevitable conclusion from this inspection ... 10
    The appendix of chapter 1:
        The fable of 'the Blind Men and the Elephant' ... 11

Chapter 2: Inspect the Necessary Pillar of the Big Bang Theory ... 13
    The prerequisite for this necessary pillar to be valid ... 13
    Inspect whether this prerequisite is satisfied or not ... 15
    The unavoidable conclusion from this inspection ... 18
    The appendix of chapter 2:
        Gamma ray bursts and ultrahigh-energy cosmic rays ... 20
            Gamma ray bursts ... 20
                The history of gamma ray bursts ... 20
                Gamma ray bursts—one of the greatest mysteries in science ... 20
                The five known features of gamma ray bursts ... 21
            Ultrahigh-energy cosmic rays ... 24

Chapter 3: Inspect the Theoretical Pillar of the Big Bang Theory ... 27
    The prerequisite for this theoretical pillar to be valid thus reliable ... 27
    Inspect whether this prerequisite is satisfied or not ... 30
    The inescapable conclusion from this inspection ... 34
    Go the extra mile ... 35

Chapter 4: The New Theory Shows Why Time Runs Slower in a Gravitational Field ... 41
    The basic principle in science ... 41
    The theoretical basis of this new theory ... 43
    The four main features of this new theory ... 45

CONTENTS

The core concept of this new theory	50
The principle of gravitational redshift in this new theory	55
The principle of gravitational light bending in this new theory	60

Chapter 5: A Fundamentally New Black Hole Theory: Mechanism-Revealed Black Hole Theory — 65

The two basic features of this new black hole theory	65
The mechanism and essence of this new black hole theory	68
The size of a mechanism-revealed black hole	71
The three fundamental natures of mechanism-revealed black holes	74
The great gravitational redshift caused by mechanism-revealed black holes	78
The gravitational light bending caused by mechanism-revealed black holes	82
The appendix of chapter 5: the event horizon of a black hole turns out to be an unreliable concept	86

Chapter 6: Unveil the Mystery of Dark Matter — 95

The concept of dark matter	95
Dark matter—one of the greatest puzzles in science	97
The mechanism, essence and definition of dark matter	101
Reveal the four fundamental natures of dark matter	103
Reveal the constituents of dark matter	103
Dark matter can emit a large amount of light	109
The great gravitational redshift caused by dark matter	110
The gravitational light bending caused by dark matter	110
The Bitter lessons after solving the problem of dark matter	112
The appendix of chapter 6:	
The Law of Mass Doing Work—about Why Mass Has Energy	114
The questions pointing to the law of mass doing work	114
The law of mass doing work	116
The mechanism of the mass-energy equivalence equation	118
The mass consumption	123

Chapter 7: After Inspecting All the Three Pillars of the Big Bang Theory — 127

GLOSSARY	133
INDEX	139

FOREWORD

The purpose of this book is to terminate the controversy over the big bang theory that has been hovering over and lingering around this theory since its birth. This purpose has been achieved through finishing a comprehensive and penetrating inspection of **all** the three pillars of this theory.

This comprehensive and penetrating inspection follows such a simple and clear principle: if all the three pillars of the big bang theory are valid, then this theory is valid thus correct and reliable; then the long-standing controversy over this theory is terminated. On the contrary, if all the three pillars of the big bang theory turn out unable to be valid (that is, if it turns out that these three pillars cannot be valid), then this theory cannot be valid thus can neither be correct nor reliable in truth (that is, this theory is invalid thus incorrect and unreliable in fact); then the long-standing controversy over this theory is also terminated.

The crucial pillar of the big bang theory is the idea of the expanding universe, which is the product of interpreting the redshift of the light radiated from the observed stars far from the earth in many other galaxies as these stars are moving away from the earth. Because this idea absolutely necessitates and totally depends on the <u>prerequisite</u> that the gravitational redshift in the universe must be insignificant, the key or focal point in this inspection is: can the **tremendous** gravitational redshift caused by the **huge** amount of <u>dark matter</u> in the universe overthrow or invalidate such a prerequisite, thus literally make this crucial pillar collapsed? Moreover, what should be noticed or reminded in this inspection is such an important or necessary clarification: at the time when the idea of the expanding universe was proposed in the early 20th century, the effect of dark matter on gravitational redshift was not taken into account, because the concept of dark matter did not get enough attention at that time. (The task of inspecting the crucial pillar of the big bang theory has been completed in chapter one.)

The necessary pillar of the big bang theory, which is the cosmic microwave background (radiation), totally depends on and hinges on the <u>prerequisite</u> that the assumed big bang was the *only* source of the cosmic microwave radiation measured nowadays. Therefore, the key or focal point in this inspection is: can such a prerequisite be satisfied, considering that the observed features of gamma ray bursts have explicitly revealed, also unavoidably displayed, such a hard *fact:* in the universe, even merely in the Milky Way galaxy, there are *numerous* celestial bodies that are the sources of gamma ray bursts at present; and these celestial bodies are also the main or important sources of the cosmic microwave radiation measured nowadays? Moreover, what should be cautioned or realized in this inspection is such an important and necessary clarification: the prevalent or conventional interpretation like 'the cosmic microwave background (radiation) is not associated with any star, galaxy, or other object' **does not include** the *numerous* celestial bodies that are the sources of gamma ray bursts, because gamma ray bursts were reported about one decade later than the cosmic microwave background (radiation). (The task of inspecting the necessary pillar of the big bang theory has been completed in chapter two.)

The theoretical pillar of the big bang theory is general relativity that is supposed to have solved the fundamental problem in front of itself, which is *why* space and time are variable thus relative in a gravitational field, or *why* time runs slower and *why* length becomes shorter in such a situation. And so, the key or focal point in this inspection is: is there sufficient and explicit evidence showing and witnessing that general relativity is literally unable to solve this fundamental problem? Moreover, is this sufficient and explicit evidence a basic, inherent and prominent feature of general relativity? (The task of inspecting the theoretical pillar of the big bang theory has been completed in chapter three.)

What chapter four has introduced is a newly developed and verified gravitational theory that has solved such a fundamentally important problem: *why* and how space and time are variable thus relative in a gravitational field (by revealing and determining *why* and how the scales of space and time are reduced in such a situation), or that has shown us *why* time runs slower and *why* length becomes shorter in a gravitational field (by revealing the mechanism behind these two *whys*). One of the most important purposes of introducing such a new gravitational theory is to provide a sharp contrast that can greatly and markedly help one see further clearly and explicitly, also much more distinctly and manifestly, whether general relativity is really unable to solve the fundamental problem in front of itself, which is *why* space and time are variable thus relative in a gravitational field, or *why* time runs slower and *why* length becomes shorter in such a situation. And so, such a sharp contrast can substantially and explicitly help one understand the related conclusion obtained after inspecting the theoretical pillar of the big bang theory.

The great gravitational redshift caused by the *numerous* and *massive* black holes in the universe (revealed and explained by a new black hole theory based on the new gravitational theory introduced in chapter four, being presented in chapter five) is substantially helpful for one to perceive and realize the gravitational redshift caused by black holes graphically and explicitly, thereby being remarkably helpful for one to understand the related conclusion obtained after inspecting the crucial pillar of the big bang theory. In addition, what one will see in chapter five is also a confirmation of the hard *fact* revealed or shown in chapter two (which is: in the universe, even merely in the Milky Way galaxy, there are *numerous* celestial bodies that are the sources of gamma ray bursts and ultrahigh-energy cosmic rays at present; and these celestial bodies are also the main or important sources of the cosmic microwave

FOREWORD

radiation measured nowadays). So this confirmation is substantially and explicitly helpful for one to understand the related conclusion obtained after inspecting the necessary pillar of the big bang theory.

The gravitational redshift caused by dark matter (revealed and explained in chapter six based on the new black hole theory introduced in chapter five) provides a substantial help for one to perceive and realize the **tremendous** gravitational redshift caused by the **huge** amount of <u>**dark matter**</u> in the universe much more specifically and explicitly, thus being considerably helpful for one to understand the related conclusion obtained after inspecting the crucial pillar of the big bang theory.

All in all, after comprehensive and penetrating inspection of <u>all</u> the three pillars of the big bang theory, the long-standing controversy over this theory is completely over. This is definitely a remarkably important event worthy to be celebrated or commemorated, considering the great importance and widespread influence of the big bang theory in science.

Bingcheng Zhao

Chapter 1

Inspect the Crucial Pillar of the Big Bang Theory

[The Window on This Chapter]

The crucial pillar of the big bang theory, which is the expanding universe (being the product of interpreting the redshift of the light radiated from the observed stars far from the earth in many other galaxies as these stars are moving away from the earth), absolutely necessitates and totally depends on the prerequisite that the gravitational redshift in the universe must be insignificant. However, it turns out that this prerequisite is not, and cannot be, satisfied at all; for instance, the **tremendous** gravitational redshift caused by the **huge** amount of **dark matter** in the universe explicitly and unavoidably shows that this prerequisite does not and cannot hold in truth. An important or necessary clarification in this inspection is: at the time when the idea of the expanding universe was proposed in the early 20th century, the effect of dark matter on gravitational redshift was not taken into account, because the concept of dark matter did not get enough attention at that time.

The prerequisite for this crucial pillar to be valid
What is the crucial pillar of the big bang theory? Answer: the crucial pillar of the big bang theory is the idea of the expanding universe. (The response from the big bang theory: yes, this answer is definitely correct, so I totally agree with it, because this idea is indeed my crucial pillar.)

Since the crucial pillar of the big bang theory is the idea of the expanding universe, whether this crucial pillar is valid or not (that is, whether this crucial pillar is able to stand its ground, or whether this crucial pillar is tenable or not) is totally dependent on and completely determined by whether this idea is valid or not. And so, the task of inspecting whether this crucial pillar is valid or not will be put into action by carrying out the task of inspecting whether this idea is valid or not.

Then how to inspect whether the idea of the expanding universe is valid or not? Because this idea originally came from the application of the Doppler effect in the 1920s (due to the American astronomer Edwin Hubble. The result gotten from this application is thus referred to as Hubble's law in professional materials. That is, the Doppler effect did play a leading or pioneering, also crucial or decisive, role for the birth of this idea), the task of inspecting whether this idea is valid or not is to be implemented by carrying out the task of inspecting whether the Doppler effect, when it is applied to the light radiated from the observed stars far from the earth in many other galaxies, can provide valid observational evidence for this idea (that is, by checking up on whether the Doppler effect is really able to provide valid evidence for this idea or not). If the answer is YES, then this idea is valid; on the contrary, if the answer is NO, then this idea is not valid or cannot be valid in truth. So the basis or rule of judging the conclusion to be drawn from this inspection is rational, objective, clear and fair. (Let us carry out this task. It's not a difficult one!)

How to carry out this task then? Answer: a simple and effective method is from the *prerequisite* for the Doppler effect to be able to provide valid observational evidence for the expanding universe. Then what is this *prerequisite?* Clearly, the ideal prerequisite is to ensure that the entire redshift of the light radiated from the observed stars is due to they are moving away from the earth. However, because the Doppler effect *on its own* is unable to tell apart whether a redshift of light is due to a star moving away from the earth or due to a gravitational field, this prerequisite thus becomes that the redshift due to gravitational fields (that is, gravitational redshift) is so small that it can be negligible, compared to the redshift due to the star moving away from the earth. That is to say, this *prerequisite* is that *gravitational redshift must be insignificant!* (Friendly reminder: such a prerequisite is also the prerequisite for the crucial pillar of the big bang theory to be valid, simply because this crucial pillar is the idea of the expanding universe, as just mentioned above.)

If such a prerequisite holds, one can come to the conclusion that the Doppler effect is able to provide valid evidence for the idea of the expanding universe; if not, one cannot draw such a conclusion of course, to be rational, objective and fair. That is, if such a prerequisite turns out unable to hold, one has no choice but to reach the unavoidable conclusion that the Doppler effect cannot provide valid observational evidence for this idea. Therefore, the task of inspecting whether the idea of the expanding universe is valid or not turns into the specific task of inspecting whether this prerequisite is satisfied or not (i.e., whether it holds or not; whether it is valid or not; whether it is true or false).

Then how to carry out this specific task? Certainly, it ought to be carried out from the present available knowledge, actually the known facts. We are going to inspect this prerequisite from four viewing angles. (Related question and answer: why are you going to inspect this prerequisite from four viewing angles, instead of simply one or two? It seems that one viewing angle is okay, and two viewing angles are enough. Answer: in order to check up on this crucial prerequisite carefully in a comprehensively and systematically thinking way, for avoiding the mistakes or for eliminating the incorrect conclusions like in the famous fable of 'the Blind Men and the Elephant'—this fable has been provided in the appendix of this chapter at its end. It seems both reasonable and safe to believe that all rational people, needless to say professional scientists, would/could naturally or easily agree that thinking comprehensively and systematically is crucially important in science: thinking in such a way or mode is definitely necessary to ensure that the obtained conclusion is objective and fair. Moreover, the history of science development has clearly told us such an irrefutable fact: only objective and fair conclusions are really valid and reliable, and only really valid and reliable conclusions can stand up to the test of time and history, especially the history of science development.)

Inspect whether this prerequisite is satisfied from four angles

In this section, we are going to inspect whether the *prerequisite* that **gravitational redshift must be insignificant** is satisfied or not from the following four viewing angles.

First angle: from the fact that the sun, even the much smaller earth, can cause significant gravitational redshift. The gravitational redshift in the gravitational field of the sun was observed for the first time by a team from Princeton University in 1962. And the gravitational redshift caused by the gravitational field of the earth was measured in 1959 in the famous Harvard tower experiment, also referred to as Pound-Rebka experiment. What should be emphasized is that this measured result is just the gravitational redshift that takes place only within the distance as short as about 20 meters very near the surface of the earth (i.e., immediately above the surface of the earth); by this emphasis, I want to point out that the total gravitational redshift across the entire gravitational field of the earth is much, much larger.

Given that the sun is just an ordinary, average-sized star that can cause significant gravitational redshift, it is clearly groundless to say that the gravitational redshift of other stars is insignificant. Given that even the much smaller earth (whose mass is far smaller than the sun; the mass of the sun is about 333,000 times that of the earth) can also cause significant gravitational redshift, clearly, even obviously, it is groundless to say that the gravitational redshift of other stars is insignificant. Therefore, the solid evidence coming from this viewing angle is sufficient to show us: the *prerequisite* that gravitational redshift must be insignificant is not satisfied in fact (that is, this prerequisite turns out to be false in truth; this prerequisite cannot hold at all). Thus, the result from this viewing angle has to say NO to such a prerequisite.

*What should be pointed out or noticed is that the very *fact*, which is that both the sun and the earth can cause significant gravitational redshift, has already become general knowledge or common sense to the related professional people in modern physics and in astronomy.

Second angle: from the location of the sun and the earth in (our) the Milky Way galaxy. The sun and the earth are largely at the edge of the Milky Way galaxy (in terms of the region occupied by an overwhelming majority of the stars in this galaxy). This feature determines that the gravitational field at the location of the sun and the earth is comparatively weaker (because the strength of the gravitational field of the Milky Way galaxy tends to *decrease* radially from its center: the closer to the center of the galaxy, the stronger the gravitational field; the farther from the center of the galaxy, the weaker the gravitational field). In fact, each of the gravitational fields of the observable galaxies has the similar tendency. It is thus rather reasonable to infer that the gravitational fields at the locations where reside many observed stars (particularly those "interpreted" as the observational evidence for the expanding universe) in many other galaxies can be much stronger than at the location of the sun and the earth. Therefore, the resulting gravitational redshift can definitely be significant. In other words, the result from this viewing angle also has to say NO to the *prerequisite* that gravitational redshift must be insignificant.

*What should be mentioned is: it is no exaggeration to say that the widely known *fact*, which is that the strength of the gravitational fields of the observable galaxies tends to *decrease* radially from their centers, has become general knowledge or common sense to the related professional people in modern physics, astronomy and cosmology (cosmology—the scientific study of the universe).

Third angle: from the angle of dark matter. While the fundamental nature of dark matter has been a thorny problem in science for more than several decades, it has been known that the existence of dark matter causes two noticeably observed gravitational effects (these two effects have been fully recognized by the related scientific communities; that is, the related professional people in the related fields of science, such as modern physics, astronomy and cosmology, have been very familiar with these two effects). One is that the stars in the outer regions of a galaxy orbit *much* faster than they would if there were only ordinary matter present. (Attention, please! The inevitable or inescapable implication of this effect is: due to the existence of dark matter, the gravitational field over the entire region of a galaxy is *much* stronger than it would be if only ordinary matter existed.)

Another effect is that the light rays from distant stars are bent (in the gravitational field of dark matter) by the enormous gravity of dark matter (such a phenomenon due to the huge gravitational effects of dark matter is referred to as the gravitational lens, being one of the main, effective and simple ways to identify the existence of dark matter nowadays). (Attention, please! The inevitable or inescapable implication of this noticeably observed gravitational effect is gravitational redshift, because of the known *fact:* light bending and gravitational redshift are the two inseparable aspects of the nature of light behaving in a gravitational field; this *fact* had been clearly shown by the results obtained from observing the behavior of light in the gravitational field of the sun in the 20th century.) Moreover, it has been known that a **huge** amount of dark matter exists in the Milky Way galaxy and other galaxies; and it is estimated that in the universe, the amount of dark matter is about **5.5 times** of that of the ordinary matter (such as the matter of stars and planets)—that is, about **85%** of the total matter or mass of the universe is dark matter. (Commentator or friendly reminder: yes, that's definitely true. As a matter of fact, the related knowledge or information, which is that the amount of dark matter is about 5.5 times of that of the ordinary matter in the universe, or about 85% of the total matter or mass of the universe is dark matter, has already become common knowledge to the related professional people in modern physics, astronomy and cosmology.)

Such a huge amount of dark matter can definitely cause significant gravitational redshift—through the inevitable or inescapable implications of the two noticeably observed gravitational effects just mentioned above (along with that the area significantly affected by a cluster of dark matter is often millions, even billions, of times the area that is occupied by the cluster of dark matter, because there is a tremendously huge extending region radially spreading out a cluster of dark matter. That is, the gravitational fields in these significantly affected regions are *much* stronger than they would be if only ordinary matter existed. And so, the light radiated from the stars in these significantly affected, huge extending regions must experience large gravitational redshift). As a result, the existence of dark matter explicitly and unavoidably shows us such a clear and definite fact: the *prerequisite* that gravitational redshift must be insignificant is not satisfied at all, and cannot be satisfied at all (that is, such a prerequisite cannot hold at all; such a prerequisite is not valid at all; such a prerequisite turns out to be false). Therefore, the result from the angle of dark matter has no choice but to say NO to the *prerequisite* that gravitational redshift must be insignificant.

*Having noticed the existence of such a **huge** amount of dark matter in the universe (again the amount of dark matter is about **5.5 times** of that of the ordinary matter; that is, about **85%** of the total matter or mass of the universe is dark matter), one can suddenly, but clearly and definitely, realize that a present opinion, like: the gravitational fields of other galaxies are not large enough to cause a significant effect on gravitational redshift, turns out to be utterly groundless in fact. (Commentator: whenever the related experts were talking about the Doppler effect as the observational evidence for the expanding universe, they should not have forgotten gravitational redshift. And whenever these experts were thinking of gravitational redshift, they should not have forgotten the **enormous** 'contribution' from the **huge** amount of dark matter in the universe. It is obviously a historic, grave and fatal mistake to have forgotten or overlooked the **tremendous** gravitational redshift caused by the **huge** amount of dark matter in the universe!)

(*Friendly reminder: for other more specific and detailed evidence and knowledge about the tremendous gravitational redshift caused by the huge amount of dark matter in the universe, please see chapter six; the evidence and knowledge presented there are substantially helpful for one to perceive, realize and understand the gravitational redshift caused by dark matter graphically thus more explicitly.)

Fourth angle: from the effects of black holes. Though the question like 'what is the fundamental nature of black holes?' has been a long-standing question in science, it has been known that the existence of a black hole, due to its highly massive feature (it's not unusual that the mass of a typical black hole can be hundreds of times that of the sun), can cause the gravitational field in the vast extending region outside it to be significantly stronger, actually much stronger, than without the black hole. (Related information: the region significantly affected by a black hole is *much, much larger* than that occupied by the black hole, often millions, even billions, of times larger, because there is a hugely vast extending region radially spreading out from a black hole.) Moreover, it is estimated that there are *millions* of black holes in the Milky Way galaxy; one can thus logically or rationally infer that a much larger number of black holes exist in many other galaxies, and that a much, much larger number of black holes, actually or practically an incalculable number of black holes, exist in the almost or virtually incalculable galaxies of the vast universe.

Not only that, it has been recognized that there is a supermassive black hole at the center of the Milky Way galaxy; in fact, at the center of each observable galaxy, there is also a supermassive black hole. What should be noticed is that the mass of each supermassive black hole of this type is in the range of millions, even billions, of times the mass of the sun. And what's more noticeable is that the existence of such a supermassive black hole can, not only significantly but also greatly, strengthen the gravitational field extensively, virtually over the entire region of the galaxy in which *the* black hole resides.

Clearly, these *numerous* and *massive* black holes can definitely cause significant, even very large, gravitational redshift in the vastly extending regions outside them: the light radiated from the stars in such regions must experience large gravitational redshift! Therefore, the available knowledge about black holes explicitly and inevitably tells us: the *prerequisite* that gravitational redshift must be insignificant is not satisfied at all, and cannot be satisfied at all (i.e., this prerequisite turns out to be false; this prerequisite cannot hold at all; this prerequisite is not valid at all). Consequently, also unavoidably, the result from the angle of black holes cannot help but say NO to the *prerequisite* that gravitational redshift must be insignificant, because this prerequisite turns out to be neither valid nor true in fact.

*Having been aware of the existence of *numerous* and *massive* black holes in the universe, one can suddenly, but clearly and surely, realize that a present opinion, like: the gravitational fields of other galaxies are not large enough to cause a significant effect on gravitational redshift, turns out to be obviously groundless in truth. (Commentator: whenever the related experts were talking about the Doppler effect as the observational evidence for the expanding universe, they should not have forgotten gravitational redshift. And whenever these experts were thinking of gravitational redshift, they should not have forgotten the **great** 'contribution' from the **numerous** and **massive** black holes in the universe. Undoubtedly, it is a grave and fatal mistake to have forgotten or overlooked the **great** gravitational redshift caused by the **numerous** and **massive** black holes in the universe!)

(*Friendly reminder: for other more specific and detailed evidence and knowledge about the great gravitational redshift caused by the numerous and massive black holes in the universe, please see chapter five; the evidence and knowledge presented there are considerably helpful for one to perceive and understand the gravitational redshift caused by black holes graphically thus more explicitly.)

All in all (and altogether), the results from the four viewing angles above not only collectively and consistently point to, but also sufficiently and explicitly show such a solid conclusion: the *prerequisite* that gravitational redshift must be insignificant is not satisfied and cannot be satisfied at all. That is, this prerequisite cannot hold at all; this prerequisite is not, and cannot be, valid at all; this prerequisite turns out to be definitely invalid; this prerequisite turns out to be obviously false.

The inevitable conclusion from this inspection

The solid conclusion above, which is the *prerequisite* that gravitational redshift must be insignificant is not, and cannot be, satisfied at all, has no choice but to tell us or point to such a clear and definite conclusion: the Doppler effect, when it is applied to the light radiated from the observed stars far from the earth in many other galaxies, turns out unable to provide valid observational evidence for the idea of the expanding universe in fact. (Related question and answer: considering the **reality** that the Doppler effect *on its own* is unable to tell apart whether a redshift of light is due to a star moving away from the earth or due to gravitational redshift—the redshift of light due to gravitational fields, so the conclusion drawn from the Doppler effect depends on how to interpret the observed results. This interpretation, in turn, hinges on the *prerequisite* that gravitational redshift must be insignificant. Then what is the real implication or inescapable consequence of this prerequisite turning out to be false? Answer: the actual existence of gravitational redshift has been mistakenly interpreted as that the observed stars are moving away from the earth!)

This clear and definite conclusion has no alternative but to tell us such an explicit conclusion: the idea of the expanding universe turns out to be invalid or cannot be valid in truth. This explicit conclusion has no choice but to reveal and show such an inevitable, also unavoidable, conclusion: the crucial pillar of the big bang theory turns out to be invalid or cannot be valid—that is, it turns out that this crucial pillar cannot stand its ground at all; this crucial pillar turns out unable to be tenable, because the idea of the expanding universe is the crucial pillar of the big bang theory, as clearly mentioned at the beginning of this chapter.

There is a simple and clear method that can make or help one comprehend the inevitable, also unavoidable, conclusion above easily and quickly. This method is to keep asking such a plain and straightforward question like: **in the eyes of the crucial pillar of the big bang theory, where does all the gravitational redshift in the**

vast universe go? This is because in the eyes of this crucial pillar, the sun, even the much smaller earth, can cause significant and observable gravitational redshift, but no other observed stars in many other galaxies can give rise to significant gravitational redshift. This is because in the eyes of this crucial pillar, no locations (where reside the observed stars) in many other galaxies over the vast universe can have stronger gravitational fields than at the location of the sun and the earth, even the gravitational field at this location has been known to be comparatively weaker. This is because in the eyes of this crucial pillar, an enormously huge amount of dark matter in the vast universe cannot cause significant gravitational redshift. This is because in the eyes of this crucial pillar, the numerous and massive black holes in the vast universe cannot cause significant gravitational redshift. In short, in the eyes of this crucial pillar, there is utterly no gravitational redshift at all in the vast universe!!! On the whole, in the eyes of this crucial pillar, all the gravitational redshift in the entire, vast universe simply evaporates!!! Then who can swallow up all the gravitational redshift in the entire, vast universe into his stomach?!? Nobody can, of course. (Commentator: science has nothing to fear but fear the wrong mode of thinking! Incorrect thinking modes or methods are always the number one enemy of science; and the number one enemy of science is always incorrect thinking modes or methods! The tortuous history of science development has clearly shown and warned us of that.)

The appendix of chapter 1
The fable of 'the Blind Men and the Elephant'

Once upon a time, there were six blind men who lived in a village in today's India. Every day they stood on the side of the road with begging as their living. They had often heard of elephants, but had never seen one, for being blind, how could they? One morning, it happened that an elephant was led down the road at which they exactly stood. When they heard that an elephant was passing by, they asked the drover to stop the beast so that they could have a "look". Of course they could not look at him with their eyes, but they thought they might learn what kind of animal he was by touching and feeling him. Then you should see they trusted their own sense of touch so much.

The first blind man happened to put his hand on the side of the elephant. So he said: "Well, well, this beast is exactly like a wall."

The second tightly grasped one of the elephant's tusks and felt it. So he said loudly: "You're quite mistaken. He's round, smooth, and sharp. He's more like a spear than anything else."

The third happened to grab the elephant's trunk. "Both of you are completely wrong. This elephant is right like a snake," he said.

The fourth opened both his arms and closed them around one of the elephant's legs. "Oh, how blind you are!" he cried. "It's very clear that he's round and tall like a tree."

The fifth man was very tall, so he caught one of the elephant's ears. "Even the blindest person must see that this elephant isn't like any of the things you name at all!" he said. "He's exactly like a huge fan."

The sixth man went forward for touching and feeling the elephant. He was old and slow, so it took him quite some time even to find the elephant. Eventually, he got hold of the beast's tail. "Oh, how silly you all are!" cried he. "The elephant isn't like a wall, or a spear, or a snake, or a tree; neither is he like a fan. Any person with eyes on head can see that he's exactly like a rope."

After the drover and the elephant left, the six men sat by the roadside all day, quarrelling about the elephant. They could not agree with one another, because each of them so surely believed that he knew what the beast looked like. It is not merely blind men who make such stupid mistakes. People who can see sometimes may act just as foolishly.

Chapter 2

Inspect the Necessary Pillar of the Big Bang Theory

[The Window on This Chapter]

The necessary pillar of the big bang theory, which is the cosmic microwave background (radiation), totally depends on and hinges on the <u>prerequisite</u> that the assumed big bang was the *only* source of the cosmic microwave radiation measured nowadays. However, it turns out that this <u>prerequisite</u> is not, and cannot be, satisfied at all, because the observed or known features of gamma ray bursts have explicitly, also unavoidably, revealed or displayed such a hard *fact:* in the universe, even merely in the Milky Way galaxy, there are *numerous* celestial bodies that are the sources of gamma ray bursts at present; and these celestial bodies are also the main or important sources of the cosmic microwave radiation measured nowadays. An important and necessary clarification in this inspection is: the prevalent or conventional interpretation like 'the cosmic microwave background (radiation) is not associated with any star, galaxy, or other object' **does not include** the *numerous* celestial bodies that are the sources of gamma ray bursts, because gamma ray bursts were reported about one decade later than the cosmic microwave background (radiation).

The prerequisite for this necessary pillar to be valid
What is the necessary pillar of the big bang theory? Answer: the necessary pillar of the big bang theory is the cosmic microwave background, also called the cosmic microwave background radiation (CMBR, for short) (which was first observed and discovered by American astronomers Arno Penzias and Robert Wilson in 1964),

because the CMBR has been interpreted as *necessary* evidence for the big bang theory. (The response from the big bang theory: yes, I totally agree with this answer, because the CMBR is indeed my necessary pillar.)

Since the necessary pillar of the big bang theory is the CMBR, in order to inspect whether this necessary pillar is valid or not, we ought to, and have to, inspect whether the CMBR is valid evidence for the big bang theory or not. If the CMBR is valid evidence, then one can come to the conclusion that this necessary pillar is valid; if not, one cannot reach such a conclusion of course, to be rational, objective and fair. That is, if the CMBR turns out unable to be valid evidence for the big bang theory, one has no choice but to reach the unavoidable conclusion that the necessary pillar of the big bang theory turns out to be invalid or cannot be valid in truth. (Commentator: the logical relationships in this paragraph are definitely rational and objective, also quite simple and clear, thus rather easy to follow.)

Then how to inspect whether the CMBR is valid evidence (for the big bang theory) or not? In order to inspect whether the CMBR is valid evidence or not, we need to be clearly aware of the *prerequisite* for it to be valid. If this prerequisite is satisfied, one can come to the conclusion that the CMBR is valid evidence for the big bang theory. On the contrary, if this *prerequisite* cannot be satisfied, then one has no choice but to face the unavoidable conclusion that the CMBR cannot be valid evidence for this theory. So the standard or rule of judging whether the CMBR is valid evidence is not only rational and objective, but also clear and fair.

Then what is the *prerequisite* for the CMBR to be valid evidence (for the big bang theory)? Clearly, even obviously, also undoubtedly and undeniably, this *prerequisite* is that the assumed big bang was the *only* source of the cosmic microwave radiation measured nowadays. This *prerequisite* is, in effect or in practice, literally equivalent to such a hypothesis or assumption: in the numerous, virtually or almost incalculable, galaxies of the vast universe, no celestial bodies can be the main or important sources, or at least can be the significant sources, of the cosmic microwave radiation at present. (Narrator or reminder: this prerequisite is also the prerequisite for the necessary pillar of the big bang theory to be valid, because this necessary pillar is the CMBR, as just mentioned above.)

Related question and answer: how could one see the prerequisite above more clearly and explicitly, thus grasp it more tangibly and definitely, thereby understand it more impressively and confidently? Answer: the following two simple and clear examples or analogies are quite helpful.

First example or analogy: let us say a farmer, Colin, has lost a cow without any mark as identification. A few days later, a person finds a lost cow on a piece of grassland several dozen miles away from Colin's home. In order to have the valid conclusion that this cow must be Colin's, the person must make certain that *only* Colin has lost a cow.

Second example or analogy: this is also a bit like that, an environmental agency is monitoring a specific chemical compound that is very harmful to our health, and a chemical factory, factory HHH—let us say, could give off this compound more than one hundred years ago. Today, the agency monitors a high concentration of this compound in the waters of a large lake. In order to get the valid conclusion that this compound was from factory HHH, the agency has to ensure that no other factories can emit this compound at present, or at least can make sure that no other factories can be the significant sources of this compound nowadays.

(The response from the big bang theory: yes, I totally agree with this prerequisite, because it is certainly rational and objective, also quite simple and clear, thus pretty easy to understand and accept.)

Inspect whether this prerequisite is satisfied or not

In this section, we are going to carry out the task of inspecting whether the prerequisite that the assumed big bang was the *only* source of the cosmic microwave radiation measured nowadays is satisfied or not. (Friendly reminder: again, this prerequisite is, in effect or in practice, literally equivalent to such a hypothesis or assumption: in the numerous, virtually or almost incalculable, galaxies of the vast universe, no celestial bodies can be the main or important sources, or at least can be the significant sources, of the cosmic microwave radiation at present.)

How to carry out this task then? The observed or known features of gamma ray bursts and ultrahigh-energy cosmic rays will help us carry out and accomplish this task (for the specific or detailed information about these known features, please see the appendix of this chapter at its end). Why? How? The answers are coming!

As an ongoing program for about half a century since the 1960s, one of the hottest research areas in physics or astrophysics is to identify the mysterious source(s) of gamma ray bursts, that is, to find out where gamma ray bursts come from. For that reason, a large number of strict observations of gamma ray bursts have been carried out over the last several decades. Almost at the same time, observing ultrahigh-energy cosmic rays has also been well implemented as an important task, because the mysterious source(s) of ultrahigh-energy cosmic rays has been fully recognized by the scientific community as one of the most fundamental mysteries in physics or astrophysics (a branch of astronomy dealing with the physical and chemical structure of the stars, planets, etc.).

Thanks to the great and persistent efforts of many scientists from different countries for a long time, (the related) scientists have already clearly known the important features of gamma ray bursts and ultrahigh-energy cosmic rays from the cumulative results of these long-term observations (for the specific and detailed knowledge or information about gamma ray bursts and ultrahigh-energy cosmic rays, as well as these important features, please go to the appendix of this chapter at its end).

Then what have the observed or known features of gamma ray bursts and ultrahigh-energy cosmic rays explicitly revealed or displayed? Answer: the observed features of gamma ray bursts have explicitly, also unavoidably, revealed or displayed such a hard *fact:* in the universe, even merely in the Milky Way galaxy, there are *numerous* celestial bodies that are the sources of gamma ray bursts at present; and these celestial bodies are also the main or important sources of the cosmic microwave radiation measured nowadays. (A closely related clarification: the prevalent opinion or conventional interpretation like 'the cosmic microwave background (radiation) is not associated with any star, galaxy, or other object' **does not include** the *numerous* celestial bodies that are the sources of gamma ray bursts, because gamma ray bursts were first reported in 1973, whereas the cosmic microwave background (radiation) was discovered in 1964.)

Furthermore, this hard *fact* fits the known features of the cosmic microwave radiation measured nowadays, especially its highly uniform and homogeneous feature on a large scale, much more accurately, much more rationally, and much more realistically than the hypothesis or conjecture that the assumed big bang was the *only* source of the cosmic microwave radiation measured nowadays; in fact, this hard *fact*, only this hard *fact*, is able to explain *why* the cosmic microwave radiation measured nowadays is so highly uniform and homogeneous on a large scale. (On the contrary, the hypothesis or conjecture that the assumed big bang was the *only* source of the cosmic microwave radiation measured nowadays is literally unable to explain *why* the cosmic microwave radiation measured nowadays is so highly uniform and homogeneous on a large scale! Moreover, this inability has been, and is, a widely admitted *reality* by the related scientific communities, because the great mystery of *why* the cosmic microwave radiation measured nowadays is so highly uniform and homogeneous on a large scale has been constantly perplexing the related scientific communities over the last several decades. In fact, this great mystery has been, is always, and will forever be a mystery within the paradigm of this hypothesis or conjecture.)

The monitored or known features of ultrahigh-energy cosmic rays have also explicitly revealed or displayed a similar, hard *fact* (that is, in the vast universe, even merely in the Milky Way galaxy, there are *numerous* celestial bodies that are the sources of ultrahigh-energy cosmic rays at present; and these celestial bodies are also the main or important sources of the cosmic microwave radiation measured nowadays). Moreover, these two hard *facts*, when viewed simultaneously, become further clear and noticeable, because gamma rays are one of the four common types of ultrahigh-energy cosmic rays.

All in all, the two hard *facts* above explicitly show, also unavoidably point to: the *prerequisite* that the assumed big bang was the *only* source of the cosmic microwave radiation measured nowadays **is not, and cannot be, satisfied** at all—that is, this prerequisite cannot hold in fact; this prerequisite turns out unable to be valid in reality; this prerequisite turns out to be false in truth.

The unavoidable conclusion from this inspection

Since the prerequisite that the assumed big bang was the *only* source of the cosmic microwave radiation measured nowadays is not, and cannot be, satisfied (i.e., this prerequisite turns out to be invalid or false), the *prerequisite* for the cosmic microwave background radiation (CMBR) to be valid evidence for the big bang theory cannot be satisfied at all. That is, the *prerequisite* for the necessary pillar of the big bang theory to be valid cannot be satisfied at all, because this necessary pillar is the CMBR, as mentioned at the beginning of this chapter. As a result, this inspection has no choice but to have such a solid, explicit and unavoidable conclusion: the necessary pillar of the big bang theory turns out unable to be valid in fact; that is, it turns out that this necessary pillar cannot be valid in truth. (Related question and answer: how could one perceive and realize this solid, explicit and unavoidable conclusion clearly, easily and effectively? Answer: a simple and clear, also straightforward, method like the following can be quite helpful. The validity of the necessary pillar of the big bang theory completely depends on and rests with the prerequisite that the assumed big bang was the *only* source of the cosmic microwave radiation measured nowadays, whereas the cosmic microwave radiation measured nowadays doesn't need such a prerequisite at all.)

Last but not least, what should be mentioned is that the solid, explicit and unavoidable conclusion above is further consolidated or strengthened by the most important feature of the model of the big bang theory, because this most important feature is that the very prerequisite, which is that the assumed big bang was the *only* source of the cosmic microwave radiation measured nowadays, is also the necessary requirement for interpreting the CMBR as the so-called accurate confirmation of Alexander Friedmann's first assumption. (Related knowledge: Friedmann's first assumption, which says that 'the universe looks identical in whichever direction we look' in terms of the CMBR being highly uniform and homogeneous feature on a large scale, is the necessary or indispensable condition for having had the model of the big bang theory.) Now, because the prerequisite for the CMBR to be valid evidence for the big bang theory cannot be satisfied at all, the CMBR turns out unable to be valid evidence for confirming this assumption either; that is, Friedmann's first assumption has not yet experienced valid confirmation in truth. Therefore, the most important feature of the model of the big bang theory does further consolidate or strengthen the solid, explicit and unavoidable conclusion above.

As to the questions like: where does the assumed big bang come from or originate from? Is the origin of the assumed big bang a truly valid thus really reliable concept? The answers to these questions will be mentioned at the end of the last section of the next chapter. These answers will be considerably and explicitly helpful for one to have a better understanding of the subjects involved in this chapter (these answers are delightfully waiting for you there!).

The appendix of chapter 2

Gamma ray bursts and ultrahigh-energy cosmic rays

Gamma ray bursts

The history of gamma ray bursts

What are gamma rays? Gamma rays are the highest energy and the shortest wavelength of electromagnetic radiation. Gamma rays are photons in the terminology of quantum physics. At present, it is widely believed that gamma rays can be generated by nuclear reactions, because gamma rays were originally discovered as an emission of radioactive substances. Physically, gamma rays are similar to X-rays in terms of high energy, but X-rays do not generate from atomic nuclei (they are produced in other ways instead, e.g., by slowing down of the motion of high-energy electrons).

Gamma ray bursts, GRBs for short, were discovered by accident in the 1960s. GRBs were first observed in 1967, quite by accident after U.S. satellites were deployed to monitor possible violations of the Nuclear Test Ban Treaty. GRBs were first reported in 1973, based on 1969-1971 observations by the Vela military satellites monitoring for nuclear explosions in verification of the Nuclear Test Ban Treaty.

GRBs can only be observed directly from space, because the atmosphere blocks gamma rays. Therefore, the current common practice is that GRBs are detected with satellites orbiting the earth (or traveling in the solar system). Briefly, GRBs are mysterious and powerful astronomical phenomenon that emits short-lived flashes of gamma rays (extremely high-energy radiation). These bursts last up to a few seconds, occur every day, and come from different parts of the sky.

Gamma ray bursts—one of the greatest mysteries in science

For about half a century since their first observation, gamma ray bursts (GRBs, for short) have been fully recognized as one of the fundamental mysteries by the related scientific communities in physics or astrophysics, which is clearly shown by, or explicitly reflected in, the several highly authoritative sources like the following. (i) "The source of gamma ray bursts" was collectively listed as one of World's 20 Greatest Unsolved Problems by more than 60 brilliant scientists—among them, 40 Nobel laureates (in the book *The World's 20 Greatest Unsolved Problems* authored by John R. Vacca). (ii) The source of GRBs has been listed as one of the 24 outstanding unsolved problems in the encyclopedia of physics with "What is the nature of these extraordinarily energetic astronomical objects?" (iii) GRBs have been summarized as one of the greatest unsolved problems in physics by the well-known physicist Vitalii L Ginzburg. (iv) "What are gamma ray bursters?" has been collected as one of the fundamentally important open questions by John Baez (a professor of mathematics at University of California, Riverside), and so on. All in all, GRBs pose one of the greatest mysteries in physics or astrophysics.

Then what is the **crux** of the great mystery of GRBs? This **crux** is: what are the sources of GRBs? Or where do GRBs come from? Or what celestial bodies produce GRBs? This **crux** thus clearly and explicitly tells us: the mystery of GRBs lies with the mystery in their sources; that is, the mystery in the sources of GRBs represents the mystery of GRBs.

The five known features of gamma ray bursts

While the sources of gamma ray bursts (GRBs, for short) have not been identified yet within the paradigm or stereotype of modern physics as well as its based astrophysics and astronomy (or within this paradigm or stereotype, the sources of GRBs have not been, cannot be, and will never be identified at all, simply because this paradigm or stereotype is the impassable barrier or obstacle to identifying the sources of GRBs), (the related) scientists have already clearly known the five main features of GRBs from the cumulative results of the long-term observations of GRBs, thanks to the great and persistent efforts of many scientists from different countries over the last several decades.

These five known features are: (i) ultrahigh-energy and extraordinarily intense; (ii) short bursting time; (iii) higher-energy gamma rays (with higher energy density) coming first in a single burst, and different GRBs releasing different amounts of energy; (iv) random directions ; (v) irregular bursting frequency. Because these known features are extremely important, let us go over them one by one in the following paragraph, specifically and concisely.

(i) Ultrahigh-energy and extraordinarily intense. In just a few seconds, the amount of energy released in a single gamma ray burst is equivalent to all the energy stored in the sun, and GRBs shine about a million trillion times as bright as the sun. Therefore, GRBs are *the most powerful explosions* in the universe. This feature is the most important, most decisive, and most crucial feature of GRBs. (ii) Short bursting time. The bursting time of GRBs is predominantly short, even though the bursting time varies greatly in different GRBs (note from the author: the feature of short bursting time is clearly reflected or embodied in the key word 'bursts'). For instance, the reported bursting durations included: "typically lasting a few seconds"; "from only a few seconds to a minute"; "from a few millisecond to about a hundred seconds"; "about 10 milliseconds to about 100 seconds" (note: 1 millisecond = 1/1000 second); "approximately 0.1 ~ 100 seconds" … and so forth. (iii) Higher-energy gamma rays (with higher energy density) coming first in a single burst, and different GRBs releasing different amounts of energy. (a) Norris found that in a single burst, gamma rays of different energies reached the earth-orbiting detectors at slightly different times, with higher-energy gamma rays arriving before lower-energy gamma rays. (b) The finding of GRB031203 rules out the idea that all gamma ray bursts have the same energy. (iv) Random directions. It has been found that GRBs are detected from entirely random directions of the sky. (v) Irregular bursting frequency. The bursting frequency of GRBs is predominantly irregular. For instance, sometimes the reported bursting frequency of GRBs was as "occur all over the sky <u>approximately one per day</u> at very large distances from the earth". And during the 1990s, the Burst and Transient Source Experiment, an instrument aboard NASA's orbiting Compton Gamma

Ray Observatory, found that bursts lasting from fractions of a second to tens or hundreds of seconds occurred <u>roughly once a day</u>. Nevertheless, it was also reported "GRBs occur <u>randomly several times a day</u> without warning". In addition, an already conclusive feature of GRBs is that GRBs are from outside the solar system.

Then what do these five main known features of GRBs clearly and surely tell us? They clearly and surely, also unavoidably or undeniably, tell us such an explicit *fact:* in the universe, even merely in the Milky Way galaxy, there are *numerous* celestial bodies that are the sources of gamma ray bursts at present; and these celestial bodies are also the main or important sources of the cosmic microwave radiation measured nowadays. (A closely related clarification: the prevalent opinion or conventional interpretation like 'the cosmic microwave background (radiation) is not associated with any star, galaxy, or other object' **does not include** the *numerous* celestial bodies that are the sources of gamma ray bursts, because gamma ray bursts were first reported in 1973, whereas the cosmic microwave background (radiation) was discovered in 1964.)

More importantly, this explicit *fact* fits the known features of the cosmic microwave radiation measured nowadays, especially its highly uniform and homogeneous feature on a large scale, much more accurately, much more rationally, and much more realistically than the hypothesis or conjecture that the assumed big bang was the *only* source of the cosmic microwave radiation measured nowadays; in fact, this explicit *fact*, only this explicit *fact*, has the ability to explain *why* the cosmic microwave radiation measured nowadays is so highly uniform and homogeneous on a large scale. (On the contrary, the hypothesis or conjecture that the assumed big bang was the *only* source of the cosmic microwave radiation measured nowadays is literally unable to explain *why* the cosmic microwave radiation measured nowadays is so highly uniform and homogeneous on a large scale! Moreover, this inability has been, and is, a widely admitted *reality* by the related scientific communities, because the great mystery of *why* the cosmic microwave radiation measured nowadays is so highly uniform and homogeneous on a large scale has been constantly perplexing the related scientific communities for a long time. In fact, this great mystery has been, is always, and will forever be a mystery within the paradigm of this hypothesis or conjecture.)

Ultrahigh-energy cosmic rays

Cosmic rays, also known as cosmic particles, are generally defined as the energetic particles originating outside of the earth. The composition of cosmic ray particles includes electrons, protons, gamma rays and atomic nuclei. Conceptually, cosmic rays can be broken into four different types: solar cosmic rays, ultrahigh-energy cosmic rays, galactic cosmic rays, and anomalous cosmic rays. Ultrahigh-energy cosmic rays primarily come from outside the solar system; for example, Victor Hess showed that the sun could not be the primary source of these cosmic rays. Galactic cosmic rays flow into our solar system from far away in the Milky Way galaxy; that is, galactic cosmic rays come from far beyond the solar system. Cosmic rays, except gamma rays (being a form of light with the shortest wavelength) that travel at the speed of light, travel at nearly the speed of light.

Quite similar to that gamma ray bursts have been fully recognized as one of the fundamental mysteries by the related scientific communities in physics or astrophysics over about half a century, ultrahigh-energy cosmic rays have also been fully recognized as one of the most fundamental mysteries by the related scientific communities in physics or astrophysics over the last several decades. This is explicitly reflected in the several highly authoritative sources like the following. (i) In physics, the long-term unsolved fundamental problem about cosmic rays is "Where do ultrahigh-energy cosmic rays come from?" For instance, this problem has been defined as one of the 125 "the most compelling puzzles and questions facing scientists today" by *Science* Magazine (the issue of July 8, 2005, to be exact; this magazine is widely known to be one of the most authoritative, most influential and most famous publications in science). And this puzzle has been specified in *Science* Magazine as that: "Above a certain level of energy, cosmic rays don't travel very far before being destroyed. So why are cosmic-ray hunters spotting such rays with no obvious source within our galaxy?" (ii) This fundamental puzzle has also been presented (under "the GZK paradox") as one of the 24 outstanding "unsolved problems in physics" in the encyclopedia of physics as that: why do some cosmic rays appear to possess energies that are theoretically too high? Cosmic rays have been observed at much higher energies than supernovae remnants can generate, and "where these ultrahigh energies come from" is a big question. (iii) 'The origin of ultrahigh-energy cosmic rays' has been listed as one of the greatest unsolved problems in physics by the well-known physicist Vitalii L Ginzburg. All in all, the mysterious sources of ultrahigh-energy cosmic rays have been widely known as one of the most fundamental mysteries in physics or astrophysics over several decades.

While the sources of ultrahigh-energy cosmic rays have not been identified yet within the paradigm or stereotype of modern physics as well as its based astrophysics and astronomy (or within this paradigm or stereotype, the sources of ultrahigh-energy cosmic rays have not been, cannot be, and will never be identified at all, simply because this paradigm or stereotype is the impassable barrier or obstacle to identifying the sources of ultrahigh-energy cosmic rays), (the related) scientists have already known the main features of ultrahigh-energy cosmic rays from the cumulative results of the long-term observations, thanks to the great and persistent efforts of many scientists from different countries over the last several decades. These main features are very similar to the five main known features of gamma ray bursts just mentioned above, because gamma rays are one of the four common types of ultrahigh-energy cosmic rays.

Then what have the known features of ultrahigh-energy cosmic rays explicitly shown us? Answer: these known features have explicitly, also unavoidably, shown us such a basic *fact:* in the vast universe, even merely in the Milky Way galaxy, there are *numerous* celestial bodies that are the sources of ultrahigh-energy cosmic rays at present; and these celestial bodies are also the main or important sources of the cosmic microwave radiation measured nowadays. (A closely related clarification: the prevalent opinion or conventional interpretation like 'the cosmic microwave background (radiation) is not associated with any star, galaxy, or other object' **does not include** the *numerous* celestial bodies that are the sources of ultrahigh-energy cosmic rays, because the cosmic microwave background (radiation) was observed and reported much earlier than ultrahigh-energy cosmic rays.)

Not only that, this basic *fact* fits the known features of the cosmic microwave radiation measured nowadays, especially its highly uniform and homogeneous feature on a large scale, much more accurately, much more rationally, and much more realistically than the hypothesis or conjecture that the assumed big bang was the *only* source of the cosmic microwave radiation measured nowadays; in fact, only this basic *fact* is able to explain *why* the cosmic microwave radiation measured nowadays is so highly uniform and homogeneous on a large scale. (On the contrary, the hypothesis or conjecture that the assumed big bang was the *only* source of the cosmic microwave radiation measured nowadays is literally unable to explain *why* the cosmic microwave radiation measured nowadays is so highly uniform and homogeneous on a large scale! Moreover, this inability has been, and is, a widely admitted *reality* by the related scientific communities, because the great mystery of *why* the cosmic microwave radiation measured nowadays is so highly uniform and homogeneous on a large scale has been constantly baffling the related scientific communities for quite a long time. In fact, this great mystery has been, is always, and will forever be a mystery within the paradigm of this hypothesis or conjecture.)

Chapter 3
Inspect the Theoretical Pillar of the Big Bang Theory

[The Window on This Chapter]

The theoretical pillar of the big bang theory is general relativity that is supposed to have solved the fundamental problem in front of itself, which is *why* space and time are variable thus relative in a gravitational field, or *why* time runs slower and *why* length becomes shorter in such a situation. However, there is sufficient and explicit evidence showing and witnessing that general relativity is literally unable to solve this fundamental problem. This hard evidence turns out to be an absolutely indispensable postulate employed by general relativity (this postulate says that the scales of length and time at different points over an entire gravitational field are the same). Why? If general relativity had been able to solve this fundamental problem, this indispensable postulate would not have been necessary at all, thus would never have appeared at all! If general relativity had been able to solve this fundamental problem, who would have looked for trouble by proposing such a totally redundant, completely unnecessary postulate?!? Therefore, the plain *truth* that general relativity absolutely and desperately necessitates this indispensable postulate is actually the sufficient and explicit evidence showing and witnessing this inability of general relativity.

The prerequisite for this theoretical pillar to be valid thus reliable
What is the theoretical pillar of the big bang theory? Answer: the theoretical pillar of the big bang theory is general relativity (also called the general theory of relativity or the theory of general relativity), which was established by Albert Einstein in the early 20th century. (The response from the big bang theory: yes, this answer is definitely

correct, so I totally agree with it, because general relativity is indeed my theoretical pillar. Moreover, all the people who are familiar with me know that general relativity is my theoretical pillar.)

Since the theoretical pillar of the big bang theory is general relativity, clearly, even obviously, also undoubtedly and irrefutably, the prerequisite for this theoretical pillar to be valid thus reliable is also the prerequisite for general relativity to be valid thus reliable.

Then what prerequisite is indispensable for ensuring that general relativity is a valid thus reliable theory? Or what is the prerequisite for general relativity to be a valid thus reliable theory? Let us find out the answer from the major subject of general relativity.

What is the major subject of general relativity then? Stated plainly, the major subject of general relativity is about time runs slower and length becomes shorter in a gravitational field. Specifically and concisely, the major subject of general relativity is about space and time are variable thus relative in a gravitational field.

Since the major subject of general relativity is about space and time are variable thus relative in a gravitational field, clearly, even obviously, also undoubtedly and undeniably, the problem of *why* space and time are variable thus relative in such a situation, or *why* time runs slower and *why* length becomes shorter in such a situation, is definitely the most fundamental problem right in front of general relativity. Therefore, the prerequisite for general relativity to be a valid thus reliable theory is that general relativity must have the ability to solve this most fundamental problem.

Commentator A: yes, such a prerequisite is definitely rational and objective, clear and fair, because of the following self-evident, clear, even obvious, basic principle or rule in science. If a theory is unable to solve the most fundamental problem in front of itself, clearly, even obviously, also undeniably, the theory cannot be fundamentally correct in fact; if a theory cannot be fundamentally correct, then the theory cannot be correct in truth; if a theory cannot be correct, clearly, even obviously, also undeniably or irrefutably, the theory can neither be valid nor reliable. Therefore, I totally agree with the prerequisite above.

Commentator B: I also completely agree with the prerequisite above, because of the clear, even obvious, reasons like the following. If anyone has the assertion or "conclusion" that claims a theory being unable to solve the most fundamental problem in front of itself as a correct theory, this most fundamental problem will definitely say NO; the fundamental nature of science will clearly say NO; the rational people in the world will impartially say NO. Moreover, if a theory, especially a highly influential theory that involves the fundamentally important topics or fields of science, which does not have the ability to solve the most fundamental problem in front of itself, were claimed as a correct theory, science would inevitably lose a rational, objective and fair criterion; the most fundamental nature and spirit of science would be fatally damaged; science would definitely be misled onto a dangerous road!

Commentator C: I also definitely agree with the prerequisite above, because it is not only rational and objective, but also simple and clear, thus quite easy to understand and accept. And I also totally agree with the comments of the two commentators above, because these comments also completely represent my voice.

(The response from the big bang theory: yes, I totally agree with the prerequisite above, simply because this prerequisite is definitely rational and objective, also quite simple and clear, thus very easy to understand and accept. I completely agree with the comments above too, because these comments are certainly rational and objective, clearly reasonable, thus definitely justifiable.)

(The response from general relativity: yes, I completely agree with the prerequisite above without any reluctance, because this prerequisite is certainly rational and objective, very simple and clear, thus quite easy to comprehend and accept. Moreover, the comments above are considerably helpful to understanding this prerequisite.)

Then what does the prerequisite (for general relativity to be a valid thus reliable theory) explicitly tell us? It explicitly tells us: if general relativity can satisfy this prerequisite, then one can come to the conclusion that the theoretical pillar of the big bang theory is valid thus reliable; if not, one cannot reach such a conclusion of course, to be rational, objective and fair. That is, if general relativity cannot satisfy this prerequisite, one has no choice but to reach the unavoidable conclusion that the theoretical pillar of the big bang theory turns out to be invalid thus unreliable in truth, or this theoretical pillar cannot be valid thus cannot be reliable in fact. (Commentator: the logical relationships in this paragraph are definitely rational and objective, also quite simple and clear, thus pretty easy to follow.)

Inspect whether this prerequisite is satisfied or not

In this section, we are going to carry out the task of inspecting whether the prerequisite for general relativity to be a valid thus reliable theory is satisfied or not. (Friendly reminder: this prerequisite, as analyzed and displayed in the section above, is that general relativity must have the ability to solve the most fundamental problem in front of itself, which is *why* space and time are variable thus relative in a gravitational field, or *why* time runs slower and *why* length becomes shorter in such a situation.)

Clearly, even obviously, also undoubtedly, this task will be carried out and accomplished through finding out the answer to the question: does general relativity have the ability to solve the most fundamental problem in front of itself, which is *why* space and time are variable thus relative in a gravitational field, or *why* time runs slower and *why* length becomes shorter in such a situation? The answer to this question is explicit and positive: general relativity doesn't and can't have the ability to solve this most fundamental problem at all; and this inability of general relativity is actually a solid *fact* in truth. Why? How? Is there sufficient and explicit evidence that is well able to show or witness the undeniable or inevitable existence of this solid *fact*? The answers are coming!

Yes, there is indeed sufficient and explicit evidence that clearly shows or impartially witnesses the undeniable or inevitable existence of this solid *fact*. This evidence turns out to be the indispensable postulate of 'invariant scales of length and time' employed by general relativity; this postulate says that the scales of length and time at different points over an entire gravitational field are the same (note: the professional people in physics, especially those in modern physics, particularly the experts on general relativity, are very familiar with this indispensable postulate). Specifically and prominently, one could clearly perceive and explicitly realize this solid *fact* through this indispensable postulate, as long as he or she views or thinks over this solid *fact* conversely like the following. If general relativity had been able to solve this most fundamental problem, this indispensable postulate would not have been necessary at all, thus would never have appeared at all. Or stated plainly, this solid *fact* will manifest itself even further thus become more noticeable if one thinks of or asks the simple and clear question like: if general relativity had been able to solve this most fundamental problem, who would have looked for trouble by proposing such a totally redundant, completely unnecessary postulate?!? Therefore, the plain *truth* that general relativity **indispensably** and **desperately** necessitates this absolutely necessary postulate can enable one to perceive this solid *fact* clearly and realize it explicitly, because this plain *truth* is actually the sufficient and explicit evidence showing or witnessing the undeniable existence of this solid *fact*.

Since this plain *truth* turns out to be actually the sufficient and explicit evidence showing or witnessing the undeniable existence of this solid *fact*, no one can deny this solid *fact*; one has no choice but to admit this solid *fact*. (Related question and answer: in addition to the sufficient and explicit evidence seen above, is there an important, clear clue that can greatly help one face and realize this solid *fact*? Answer: yes, there is. One of such clues is such a closely related *reality:* the problem of *why* time runs slower in a gravitational field has been perplexing many brilliant and curious physicists for a long time—for about one century since the birth of general relativity, because general relativity doesn't provide a mechanism to explain this *why*. This perplexity actually indicates that many brilliant and curious physicists have already realized that general relativity is unable to solve the problem of *why* time runs slower in a gravitational field. Quite obviously, this closely related *reality*, because it is consistent with the solid *fact* above, can considerably help one face and realize this solid *fact*.) (Commentator or reminder: this solid *fact* has no choice but to tell us such an inescapable and undeniable *reality:* while it's true that general relativity does tell us that space and time are variable thus relative in a gravitational field, it is really unable to tell us *why* space and time are variable thus relative in such a situation, simply because it is incapable of revealing the mechanism behind this *why*. That is to say, while general relativity does tell us time runs slower and length becomes shorter in a gravitational field, it is really unable to tell us *why* time runs slower and *why* length becomes shorter in such a situation, because it is incapable of revealing the mechanism behind these two *whys*.)

(The response from general relativity: I entirely agree with the solid fact that I am indeed unable to solve the most fundamental problem in front of myself, which is *why* space and time are variable thus relative in a gravitational field, or *why* time runs slower and *why* length becomes shorter in such a situation, because I have been completely and genuinely convinced by the analyses above. As a matter of fact, in my indispensable postulate of 'invariant scales of length and time', through the word 'postulate', I have honestly, though somewhat tacitly, admitted that I don't know the *whys* above, because the purpose or task of *any postulates* is not to deal with the questions or problems about *whys* at all. In addition, the above analyses are not only undoubtedly rational and objective, but also quite simple and clear, thus pretty easy to understand and accept.)

(The response from the big bang theory: I also completely agree with the solid fact that general relativity is unable to solve the most fundamental problem in front of itself, because I have clearly seen that the indispensable postulate of 'invariant scales of length and time' employed by general relativity is really and truly the sufficient and explicit evidence showing or witnessing the very existence of this solid fact; because I cannot find out any valid reasons to deny the existence of this solid fact. That is to say, I have no choice but to admit that my theoretical pillar turns out to be a theory that doesn't have the ability to solve the most fundamental problem right in front of itself.)

Even though the solid *fact* that general relativity is unable to solve the most fundamental problem in front of itself (which is *why* space and time are variable thus relative in a gravitational field, or *why* time runs slower and *why* length becomes shorter in such a situation) has been clearly shown by the sufficient and explicit evidence presented above, some careful readers may still feel perplexed on seeing this solid *fact*. These readers may have the perplexity like: it is often said that general relativity has passed observational tests—such as the rotation of the long axis of Mercury's orbit, light bending around the sun, and time running slower in the gravitational field of the earth, why does such a solid *fact* still exist? This kind of possible perplexity seems to call for a closely related clarification, which is presented as follows.

This clarification is to be completed through knowing or reviewing the pertinent, also crucially or remarkably important, general knowledge or common sense in science, which is that observational or experimental tests themselves have neither the function nor the ability to answer the questions about *whys* or solve the problems about *whys*. And so, the task or purpose of observational or experimental tests is not to deal with these questions or these problems at all; the task of answering the questions about *whys* or solving the problems about *whys* is, or is supposed to be, responsible by theories instead. Quite obviously, also rather rationally, such general knowledge or common sense clearly shows or explicitly points to: none of the observational tests of general relativity has the ability to change this solid *fact* at all, not to mention that none of these tests has ever targeted to deal with this very *fact* at all; not to mention that none of these tests has even intended to touch this very *fact!* Therefore, all these observational tests actually have nothing to do with this solid *fact* at all; that is, none of these observational tests can change or affect this solid *fact*.

(*Friendly reminder: one will see this solid *fact* even more clearly and explicitly through a sharp contrast against the new theory that has solved the fundamentally important problem of *why* and how space and time are variable thus relative in a gravitational field by revealing *why* and how the scales of space and time are reduced in such a situation, or has shown us *why* time runs slower and *why* length becomes shorter in a gravitational field by revealing the mechanism behind these two *whys*. This new theory will be introduced in the next chapter.) (Narrator: this new theory is pleasantly and zealously waiting for us; it is waving its warm greetings to all of us! This new theory is always very hospitable to all visitors and readers.)

All in all, the inspection carried out in this section has sufficiently and explicitly shown or displayed, also unavoidably pointed to such a solid *fact:* general relativity is really and truly unable to solve the most fundamental problem in front of itself: which is *why* space and time are variable thus relative in a gravitational field, or *why* time runs slower and *why* length becomes shorter in such a situation. This solid *fact* has no choice but to tell us such an inevitable and inescapable *truth:* the prerequisite for general relativity to be a valid thus reliable theory is not, and cannot be, satisfied at all.

The inescapable conclusion from this inspection

Since the prerequisite for general relativity to be a valid thus reliable theory is not, and cannot be, satisfied (i.e., it turns out that this prerequisite doesn't and can't hold at all, or this prerequisite turns out to be invalid or false), the *prerequisite* for the theoretical pillar of the big bang theory to be valid thus reliable **cannot be satisfied** at all, because this theoretical pillar is general relativity, as clearly pointed out at the beginning of this chapter. As a result—not only as a clear and definite result, but also as an undeniable and unavoidable result, this inspection has no choice but to have such a solid, explicit and inescapable conclusion: the theoretical pillar of the big bang theory turns out unable to be valid in truth; that is, this theoretical pillar is not, and cannot be, valid at all; consequently, also unavoidably, the theoretical pillar of the big bang theory turns out unable to be reliable in truth, that is, this theoretical pillar is not, and cannot be, reliable at all.

(Narrator: after seeing the solid, explicit and inescapable conclusion above, the big bang theory thinks it over seriously and rationally, ponders it over seriously and rationally, and contemplates it seriously and rationally. After that, the big bang theory completely comprehends and totally accepts this conclusion rationally, though somewhat uncomfortably emotionally.)

Go the extra mile

After seeing the solid *fact* that general relativity is indeed unable to solve the most fundamental problem in front of itself (which is *why* space and time are variable thus relative in a gravitational field, or *why* time runs slower and *why* length becomes shorter in such a situation), one might/could naturally or easily think of or raise the related question like: has this solid *fact* incurred or caused other serious consequences that are closely related to the big bang theory? The answer to this kind of question is explicit and positive: yes, it has; one of these consequences is the concept of singularity. (Related question and answer: why do you ascribe the concept of singularity to this solid *fact?* Answer: for two clear reasons. One is that singularity comes from a solution of the core equation of general relativity. Another is that the new theory, which has solved the very problem of *why* space and time are variable thus relative in a gravitational field or has revealed *why* time runs slower and *why* length becomes shorter in such a situation, shows that there is no singularity at all. This new theory is provided in the next chapter.)

Then what is singularity? What is the most important and obvious characteristic of singularity? Let us explore the answers to these questions in the following several paragraphs.

The so-called singularity, being predicted or created by general relativity, is a mathematical point whose size is zero or infinitely close to zero and *with infinite density and infinite temperature*. However, believe it or not, physically the two aspects of the concept of singularity, which are *'infinite density and infinite temperature'*, are actually incompatible thus clearly self-contradictory in essence. Specifically, please carefully notice that these two aspects cannot coexist at all: infinite density denies motion, whereas infinite temperature requires a motion at nearly the speed of light. As a result—as an explicit and undeniable result in truth, **the so-called singularity turns out to be a clearly and seriously self-contradictory concept**, being the most important and obvious characteristic of singularity.

What should be pointed out is that this most important and obvious characteristic cannot be changed or affected by any sophistry as well as the argument or opinion based on sophistry. Some readers might have already noticed that, in order to "defend" the concept of singularity, has appeared a sort of sophistry as well as the argument or opinion based on sophistry, like: one cannot really argue with a mathematical point. The opinion of this sort, however, cannot be really valid at all! Why? Clearly, when a mathematical point is used to describe the phenomena in physics, it has to obey the rules of physics. Thus, this sort of opinion, being merely a fabrication or product of covertly changing concepts or switching subjects, is utterly invalid, thereby totally futile in "defending" the concept of singularity. (Then what is a vivid example of changing concepts or switching subjects? Please allow me to show it. In a public lecture on the importance of good nutrition to our human beings, after showing various observational results, the lecturer concluded in the end: "Therefore, for a specific individual, the better nutrition one has, the stronger he or she is." A person in the audience disagreed immediately with: "The forage of horses is much less nutritious than the food of people, but horses are much stronger than people.") In addition, it should be cautioned that even the argument, which claims or views singularity as *a mathematical point,* still cannot be a truly valid excuse for "defending" the concept of singularity.

Why? Mathematics never allows two rules that are clearly and seriously self-contradictory to coexist. For example, mathematics never permits 'two plus three equals five' and 'two plus three equals six' to exist together.

More than that, the argument like 'one cannot really argue with a mathematical point' is actually unable to be truly valid for "defending" the concept of singularity, *even from the angle of singularity*. Why? Answer: because general relativity itself breaks down at singularity; that is, the concept of singularity explicitly and unavoidably tells us: general relativity has created a place (singularity) where general relativity itself is no longer workable or applicable. What should be mentioned is that the basic feature, which is 'general relativity itself breaks down at singularity', is a fully recognized feature of singularity by the related scientific communities; that is, this basic feature is a well-known, undeniable, also inescapable or unavoidable, feature of singularity.

All in all, any sophistry, including the argument or opinion based on sophistry, doesn't and can't change or affect the irrefutable *fact* that **the so-called singularity is actually a clearly and seriously self-contradictory concept**.

Interestingly or somewhat surprisingly, the most important and obvious characteristic of singularity mentioned above (which is that the so-called singularity turns out to be a clearly and seriously self-contradictory concept) can bring us something more important than this characteristic itself. What seems to be ironically true is that, while the sophistry, being represented with 'one cannot really argue with a mathematical point' analyzed and discussed above (including the argument or opinion based on this sort of sophistry), doesn't and can't change the simple *fact* that **the concept of singularity is clearly and seriously self-contradictory**, it does teach us a gravely bitter lesson (which can also be regarded either as a profound implication or as a constructive inspiration from the long-term viewpoint of science advancement).

This gravely bitter lesson is: in science, when a theory has incurred a clearly and seriously self-contradictory concept, a rational approach (or one of the rational approaches) should carefully examine whether the theory has solved the most fundamental problem right in front of itself and whether the theory has the ability to solve this most fundamental problem, rather than keep looking for various excuses or "interpretations" for such a concept. There are no valid excuses or "interpretations" at all for any clearly and seriously self-contradictory concept! Any effort or action of looking for any excuse or "interpretation" for a clearly and seriously self-contradictory concept is no less than looking for a pretext for an assertion like a 2-kilogram hen produced a 5-kilogram egg. The history of science development has clearly told us: incorrect thinking is the number one enemy of science; correct thinking is the number one friend of science!

Having known or realized the irrefutable *fact* that **the concept of singularity turns out to be indeed clearly and seriously self-contradictory** (because the two crucial, indispensable and inseparable aspects of singularity, '*infinite density and infinite temperature*', cannot coexist at all), one may think of or raise the related question like: is this irrefutable *fact* closely related to the big bang theory? The answer to this question is explicit and positive: yes, this irrefutable *fact* is indeed closely related to the big bang theory, simply because the assumed big bang originates from the presumed big bang singularity.

Then what does this irrefutable *fact* actually and unavoidably tell us regarding the big bang theory? Does this irrefutable *fact* have profound, serious, even fatal, impacts on the big bang theory? Let us analyze and discuss them through the following two sides.

On one side, since the two crucial, indispensable and inseparable aspects of the presumed big bang singularity, '*infinite density and infinite temperature*', couldn't coexist at all (because *infinite density* denied motion, whereas *infinite temperature* required a motion at nearly the speed of light), it thus becomes rational and reasonable believing that the presumed big bang singularity couldn't exist (in other words, it is actually irrational and unreasonable believing that the presumed big bang singularity could exist. Or at least it becomes much more rational and reasonable believing that the presumed big bang singularity couldn't exist than believing it could).

On the other side, as analyzed and unveiled in chapter two, the observed features of gamma ray bursts have explicitly revealed, also unavoidably displayed, such a hard *fact:* in the universe, even merely in the Milky Way galaxy, there are *numerous* celestial bodies that are the sources of gamma ray bursts at present; and these celestial bodies are also the main or important sources of the cosmic microwave radiation measured nowadays. Moreover, this hard *fact* fits the known features of the cosmic microwave radiation measured nowadays, especially its highly uniform and homogeneous feature on a large scale, much more accurately, much more rationally, and much more realistically than the hypothesis or conjecture that the assumed big bang was the *only* source of the cosmic microwave radiation measured nowadays; in fact, this hard *fact*, only this hard *fact*, is able to explain *why* the cosmic microwave radiation measured nowadays is so highly uniform and homogeneous on a large scale. (On the contrary, the hypothesis or conjecture that the assumed big bang was the *only* source of the cosmic microwave radiation measured nowadays is literally unable to explain *why* the cosmic microwave radiation measured nowadays is so highly uniform and homogeneous on a large scale! In fact, this inability has been, and is, a widely admitted *reality* by the related scientific communities, because the great mystery of *why* the cosmic microwave radiation measured nowadays is so highly uniform and homogeneous on a large scale has been constantly perplexing the related scientific communities for a long time. Moreover, this great mystery has been, is always, and will forever be a mystery within the paradigm of this hypothesis or conjecture.)

When these two sides come together, and are considered simultaneously, it is quite rational and reasonable that some, even many, attentive and insightful readers might suddenly get the realization like the following. Most probably, the assumed big bang could be merely a mistaken coincidence or "combination": general relativity, due to its inability to solve the most fundamental problem in front of itself, creates the actually not existed big bang singularity; the cosmic microwave radiation measured nowadays has been mistakenly interpreted as the evidence for this assumed big bang. (The response from me: definitely and surely, such a realization is not only quite rational and reasonable, but also pretty insightful and sharp, because the probability or chance of such a mistaken coincidence could be very high, being indicated by the combined or collective effect of the pieces of information analyzed or mentioned in the two paragraphs above. Moreover, such an insightful and sharp realization can substantially help one understand the subjects covered in chapters two and three comprehensively and effectively. So all the readers who have gotten such a helpful realization should be happy for themselves; I am also genuinely happy for them.)

Finally, what should be mentioned is that the subjects involved in this section have also provided the answers for the two remaining questions at the end of the last chapter, which are: 'where does the assumed big bang come from or originate from? Is the origin of the assumed big bang a truly valid thus really reliable concept?' Then what are these answers? They are: the assumed big bang comes from or originates from the presumed big bang singularity that is fabricated or created by general relativity (due to its inability to solve the most fundamental problem in front of itself). The origin of the assumed big bang, which is the presumed big bang singularity, is not, and cannot be, a truly valid thus really reliable concept, because this origin is simply the consequence of general relativity being unable to solve the most fundamental problem in front of itself (which is *why* space and time are variable thus relative in a gravitational field, or *why* time runs slower and *why* length becomes shorter in such a situation); because the presumed big bang singularity turns out to be a clearly and seriously self-contradictory concept, either in essence or in truth or in both.

Chapter 4

The New Theory Shows Why Time Runs Slower in a Gravitational Field

[The Window on This Chapter]

The fundamental question of '*why* time runs slower in a gravitational field' has been at last answered, and answered by the newly developed and verified gravitational theory (which is mechanism-revealed gravitational theory, the leading role of this chapter).

This new gravitational theory is the *only* scientific theory that shows us *why* time runs slower and *why* length becomes shorter in a gravitational field by revealing the mechanism behind these two *whys*.

This new gravitational theory is the fundamental, also indispensable, theoretical basis of the very theory that has the ability to unveil the mystery of dark matter (this mystery includes the constituents and fundamental nature of dark matter).

Without this new gravitational theory, it would have been definitely impossible to unveil the mystery of dark matter—if there had not been this new gravitational theory, the mystery of dark matter would have been kept mysterious forever; dark matter would thus have been in the darkness forever (fortunately, this new gravitational theory has not only already existed, but has also been comprehensively verified).

This chapter is going to introduce the (newly developed and comprehensively verified) gravitational theory that has solved the fundamentally important problem of *why* and how space and time are variable thus relative in a gravitational field (by revealing and determining *why* and how the scales of space and time are reduced in such a situation), or that has shown us *why* time runs slower and *why* length becomes shorter in a gravitational field (by revealing the mechanism behind these two *whys*). One of the most important purposes of introducing this new gravitational theory is to provide a sharp and noticeable contrast through which one could see further clearly and explicitly, actually much more clearly and explicitly, the solid *fact* mentioned in chapter three (this solid *fact* is that general relativity doesn't have the ability to solve the most fundamental problem in front of itself, which is *why* space and time are variable thus relative in a gravitational field, or *why* time runs slower and *why* length becomes shorter in such a situation).

Not only that, this new gravitational theory is the only scientific theory that has revealed the mechanism and essence of gravitational redshift, also being the only scientific theory that is able to do that. Moreover, this mechanism and essence is the key and prerequisite to having a profound, comprehensive and fundamental understanding of the gravitational redshift caused by black holes and dark matter. As a result, this new gravitational theory is not only a substantial help for one to have a better understanding of the gravitational redshift caused by black holes and dark matter mentioned in chapter one, but also the indispensable theoretical basis of the related subjects covered in chapters five and six.

Before going to the specific information of this new theory, let us review a basic principle in science, because this basic principle, being a very basic attribute of any theories in science, is fundamentally important and crucially indispensable to maintaining or upholding the integrity of science.

The basic principle in science

In science, there is such a generally acknowledged and totally accepted basic principle, which is also an objective and rational criterion: for any theory, the things that really matter lie in *what* rather than who—lie with *what* the theory talks about, instead of who developed it. What should be noticed or mentioned is that this basic principle has been completely recognized and admitted by the scientific community as general knowledge or common sense in science nowadays; thus, it is reasonable and realistic to believe that all today's scientists know this basic principle pretty well. On the contrary, if this basic principle were thrown away, science would inevitably lose a rational, objective and fair criterion; the most fundamental nature and spirit of science would be fatally damaged; science would definitely be misled onto a dangerous track; science would no longer be science at all! Therefore, this basic principle, because it is fundamentally important and crucially indispensable to the development and advancement of science, is indeed the cornerstone of science or can play the role quite similar to the cornerstones of a building.

Correspondingly, in order to maintain or uphold the integrity of science through ensuring its objectivity, truthfulness and reliability, there is such a self-evident rule, which is also a very basic professional requirement for anyone to introduce a theory, no matter who developed it: he or she is NOT allowed to make any exaggerated descriptions about the theory—including its meanings, implications and significance. And more specifically to be further explicit, he or she must only tell truth and fact; he/she must ensure that only truth and fact are introduced or presented. The response or promise from me: surely and of course, I will strictly obey this self-evident rule and abide by this very basic professional requirement in introducing any theories, including new theories!

The theoretical basis of this new theory

The theoretical basis of this new theory is mechanism-revealed scales relativity theory (MRSRT, for short) that unveils *why* time runs slower and *why* length becomes shorter at high speed by revealing the mechanism behind these two *whys*. MRSRT is of mechanism-revealed nature, because its theoretical foundation is of mechanism-revealed nature; because it reveals the mechanism behind its describing phenomena; and because it uncovers and determines *why* and *how* mass, length and time are variable thus relative at different speeds by finding out the relationship in their scales.

MRSRT has the following unprecedented, fundamentally important, also historic, great functions. (i) It unveils *why* time runs slower and *why* length becomes shorter at high speed by revealing the mechanism behind these two *whys*, or it shows and determines *why* and *how* the scales of time and length are reduced at high speed. (ii) It reveals the mechanism behind the first postulate of special relativity (this postulate says that the speed of light is the same for all observers, regardless of their motion relative to the source of light, or being simply referred to as 'the constancy of the speed of light' or as 'the constant speed of light'. Thus, the revealing of the mechanism behind this postulate finally answers the long-standing, also fundamentally important, exceptionally famous question of *why* the speed of light is constant). (iii) It reveals the mechanism behind the second postulate of special relativity (this postulate says that all observers moving at constant speed should have the same physical laws; that is, the laws of science should be the same for all freely moving observers, no matter what their speed. So the revealing of the mechanism behind this postulate finally finds out the substantially important *why* behind or underlying this highly influential postulate). Moreover, MRSRT has been comprehensively verified or confirmed from three different aspects or angles.

The take home message about MRSRT is: MRSRT has unveiled *why* time runs slower and *why* length becomes shorter in the situation of high speed by revealing the mechanism behind these two *whys;* that is, MRSRT has solved the long-standing, fundamentally important problems of *why* time runs slower and *why* length becomes shorter at high speed. (Friendly reminder: if one wants to know the specific and detailed knowledge or information about MRSRT, please refer to chapter two with chapter title as "Why Time Runs Slower at High Speed" (P. 13 ~ 30) in the book *The Long Night Is Over: The Mystery of Dark Matter Has Been Unveiled; The Constituents of Dark Matter Have Been Revealed.* I am concerned it would seriously distract one's attention on the new gravitational theory to be introduced in this chapter if I dwelt on MRSRT too much here.)

The four main features of this new theory

Let us start from briefly previewing the four main features of the new gravitational theory that unveils the mystery of *why* time runs slower and *why* length becomes shorter in a gravitational field by revealing the mechanism behind these two *whys* (this new theory was developed and verified by me not very long ago).

First of all and most all—first feature: this new theory solves the fundamentally important problem of *why* and how space and time are variable thus relative in a gravitational field by revealing *why* and how their scales are variable thus relative in such a situation. (That is to say, this new theory shows us *why* time runs slower and *why* length becomes shorter in a gravitational field by revealing the mechanism behind these two *whys*.) Such a fundamental feature, being the most prominent mark and highlight of this theory, is the heart and soul of this theory. (Commentator: this feature shows that this new theory has solved what general relativity cannot solve, thus reflecting the most fundamental and noticeable difference between these two theories, thereby providing a sharp contrast that is very helpful for one to see more clearly the solid *fact* that general relativity is indeed unable to solve the most fundamental problem in front of itself. So this feature is actually a good indication of the great importance and obvious necessity to introduce this new theory.)

Second feature: the predictions of this new theory accurately fit all the important observational results from different aspects. For instance, the values calculated with this theory precisely agree with the three famous observational results: the rotation of the long axis of Mercury's orbit, light bending around the sun, and the test of time running slower in the gravitational field of the earth by the Harvard tower experiment. (Commentator: such a feature answers one of the most important questions: has this theory been verified or tested? At the same time, such a feature is also a specific demonstration that this new theory, even though much simpler than general relativity, has accurately predicted what general relativity could predict.) (Reminder: as a result, this new gravitational theory has been comprehensively verified or confirmed from six different aspects altogether, because its theoretical basis, which is the new theory that reveals *why* time runs slower and *why* length becomes shorter at high speed, has also been verified or confirmed from three different aspects, as mentioned in the section above. With such comprehensive verifications or confirmations, it seems not only rather rational but also quite reasonable that one ought to be fully, at least highly, confident in this new theory, even though it is a radically new theory.)

Third feature: this new theory is quite easy to comprehend. For example, its core point, being the epitome of its core concept to be presented soon, can be easily explained via the simple and clear method like the following. Let us say, initially at a certain location far away from the gravitational field of a massive celestial body, there are two exactly identical clocks, A and B; their faces are with the same size, which means they have the same time scale. Clock A then moves to a location near the celestial body under the action of its gravity (attraction)—this motion causes the time scale of clock A to have become smaller, as revealed by the core concept of this new theory to be seen immediately, whereas clock B still stays at its original location. Now, if you read clock A with the time scale on clock B, you will find that time runs slower (that is, clock A runs slower than clock B), because clock B has a larger time scale than clock A. This is a bit like: if you read a clock of 10-centimeter face according to the time scale on a clock of 15-centimeter face, you will find that the clock of 10-centimeter face runs slower (than the clock of 15-centimeter face; you can draw two such clocks on two transparent papers, and read them). (Related question and answer: why is this new theory so easy to understand? Answer: because of its attribute, being the fourth feature of this new theory to be seen immediately.)

The New Theory Shows Why Time Runs Slower in a Gravitational Field

Fourth feature: the attribute of this new theory is of mechanism-revealed nature, being determined by the following two sufficient and explicit reasons. One is that the theoretical basis of this new theory is of mechanism-revealed nature; this theoretical basis is mechanism-revealed scales relativity theory (which has been concisely mentioned in the section above) that unveils *why* time runs slower and *why* length becomes shorter at high speed by revealing the mechanism behind these two *whys*. Another is that this new theory reveals the mechanism behind its describing phenomena; for instance, this new theory has solved the fundamentally important problem of *why* and how space and time are variable thus relative in a gravitational field, by revealing the mechanism of *why* and how the scales of space and time are variable thus relative due to the existence of the gravitational field (that is, this new theory has shown us *why* time runs slower and *why* length becomes shorter in a gravitational field by revealing the mechanism behind these two *whys*). Because of these two sufficient, explicit and necessary reasons, this new theory has been named (by me) as mechanism-revealed gravitational theory (MRGT, for short). (Related question and answer: why do you point out or mention the attribute of this new theory here? Answer: in order to provide a sharp contrast against the attribute of general relativity, being the very topic or focus to be seen immediately.)

What should be pointed out or noticed is that the attribute of this new theory, its *mechanism-revealed* nature, is **fundamentally different** from that of general relativity, because the attribute of general relativity is of *postulate-based* nature, being explicitly and sufficiently determined by the plain *fact* that general relativity is based on a set of assumptions, hypotheses and postulates (AHPs, for short). Specifically, general relativity is based on five AHPs. (1) The postulate of 'equivalence principle' says that gravitational force has the same effect in increasing the velocity of an object as other traditional forces. This postulate is the heart and soul of general relativity. (2) The postulate of 'invariant scales of length and time' says that the scales of length and time at different points over an entire gravitational field are the same. This postulate, simply stated as 'scale invariant general relativity' by Fernando Franco, is indispensable for general relativity to deal with the issues of space and time in gravitational fields. (3) The postulate of 'photons have inertial mass' is derived from the observation that photons have momentum (a quantity of motion of a moving object), by ascribing the momentum of photons to their assumed inertial mass. This postulate is crucial for general relativity to describe the behaviors of photons (light) in gravitational fields. (4) The assumption of 'the equivalence of inertial and gravitational mass' says that the gravitational mass and inertial mass of the same object are equal to each other (the inertial mass of an object is defined based on Newton's second law. This law tells us that an object with a certain amount of mass will accelerate, or change its speed, at a rate that is proportional to the magnitude or value of the net force acting on the object; this 'a certain amount of mass' is defined as the inertial mass of the object). (5) The assumption of 'the equivalence of inertial mass and active gravitational mass' is for making gravitational mass generate the curvature of space-time described in general relativity. Therefore, one can clearly see that general relativity hires at least five AHPs indeed. And so, the real or accurate name of general relativity is (or is supposed to be) ***postulate-based* general relativity**. (The response from general relativity: yes, I completely agree that my real or accurate name is supposed to be *postulate-based* general relativity, because such a name perfectly fits the crystal clear fact that I am indeed a postulate-based theory.)

*Related questions and answers: what is **the most fundamental difference** between *mechanism-revealed* gravitational theory and *postulate-based* general relativity? Answer: mechanism-revealed gravitational theory has solved the fundamentally important problem of *why* and how space and time are variable thus relative in a gravitational field, by revealing the mechanism of *why* and how the scales of space and time are variable thus relative due to the existence of the gravitational field (that is, mechanism-revealed gravitational theory has shown us *why* time runs slower and *why* length becomes shorter in a gravitational field by revealing the mechanism behind these two *whys*). Postulate-based general relativity is unable to solve the most fundamental problem in front of itself: *why* space and time are variable thus relative in a gravitational field, simply because it is incapable of revealing the mechanism behind this *why;* consequently, also unavoidably and undeniably, postulate-based general relativity doesn't and can't have the ability to solve the fundamentally important problem of *why* and how space and time are variable thus relative in a gravitational field (that is, postulate-based general relativity doesn't and can't have the ability to tell us *why* time runs slower and *why* length becomes shorter in a gravitational field). Question: does **this most fundamental difference** have or cause fundamentally profound, crucially important, also exceptionally impressive, great influences or consequences? Answer: **yes, it does**. One of these influences or consequences is witnessed by **dark matter**. Why? How? Most concisely, dear readers will clearly see in the next two chapters: mechanism-revealed gravitational theory is the most fundamental prerequisite, also the key and indispensable theoretical basis, to solve the problem of **dark matter** or unveil its mystery—this mystery includes the constituents and fundamental nature of **dark matter**; whereas postulate-based general relativity turns out to be the shackles and obstacles to solving the problem of **dark matter** or unveiling its mystery. What should be mentioned or reminded is that the problem of **dark matter** or its mystery has become one of the two long-term unsolved, greatest problems in science within the paradigm or stereotype of postulate-based general relativity; the other one is **dark energy**.

*Commentator: the four core features previewed above indicate that this newly discovered and verified gravitational theory is indeed a fundamentally and crucially important theory in science. The development of science in the 21st century, especially the development and advancement in physics, astronomy and cosmology (the scientific study of the universe), can and will witness the profound implications and great significance of this newly discovered theory.

The core concept of this new theory

Having concisely previewed the four main features of this new gravitational theory, let us go to its core concept—the representative or spotlight of this new theory. Overall, the core concept of this new theory is: the space scale and time scale in a gravitational field (being referred to as the gravitational scales of space and time) are expressed with the gravitational scale contour lines of space and time, a bit like the contour lines on a topographical map (Fig. 4.1, next page). And the value of these gravitational scale contour lines is determined (calculated) by the equation of gravitational scales of space and time, the core equation of this new gravitational theory.

Specifically, this core concept reveals the two basic aspects of the gravitational scales of space and time, or the two basic aspects of the gravitational scale contour lines of space and time (Fig. 4.1). One is that the gravitational scale contour lines of space and time in the gravitational field of a massive celestial body (the sun or the earth, for example) consist of an infinite number of concentric spheres, with the mass center of the body as their common center. Another is that the value of these gravitational scale contour lines increases radially outward across different contour lines—first rapidly then slowly, somewhat like the contour lines of a normal basin on a topographical map, or similar to the shape of the bell part of a trumpet when it is held vertically with its bell part facing upwards.

The New Theory Shows Why Time Runs Slower in a Gravitational Field

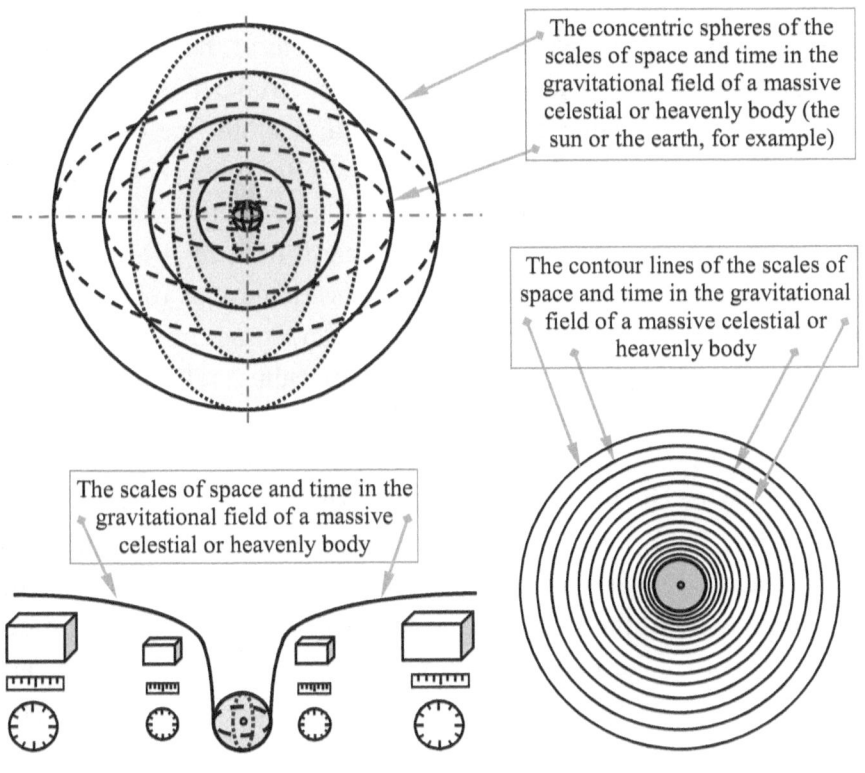

Figure 4.1, an illustration of the core result of the new gravitational theory (mechanism-revealed gravitational theory): why space and time are variable thus relative in a gravitational field (or why and how the scales of space and time are reduced in a gravitational field). So this new theory reveals the secrets of why time runs slower and why length becomes shorter in a gravitational field (in the eyes of the observers far away from the gravitational field). This new gravitational theory has been comprehensively verified or confirmed from multiple aspects or angles.

One can clearly visualize and easily grasp the above two basic aspects via the following simple and clear examples. While a plane is taking off, it is passing through the different gravitational scale contour lines of space and time in the gravitational field of the earth from low to high value. While a plane is flying at a certain fixed height, it is traveling along the same gravitational scale contour line of space and time in the gravitational field of the earth. While a plane is landing, it is passing through the different gravitational scale contour lines of space and time in the gravitational field of the earth from high to low value. (Related questions and answers: what are variable thus relative in this new gravitational theory? Answer: space and time are variable thus relative. Question: why are they variable thus relative? Answer: because their scales are variable thus relative. Question: why? Answer: because the essence of the core concept of this new theory shows so. Question: what is this essence then? Answer: the existence of a gravitational field causes a certain reduction in the scales of space and time in the gravitational field, along with that: the stronger the gravitational field, the larger the reduction. Question: how could one grasp the core concept of this newly developed theory clearly and easily, tangibly and impressively? Answer: please see the specific information in the following two paragraphs.)

The New Theory Shows Why Time Runs Slower in a Gravitational Field

One can tangibly grasp the core concept of this new theory via a simple and clear way like the following. Initially, at a location very far from the gravitational field of a massive celestial body, there are two exactly identical clocks—their faces are accurately the same size with inner diameter as one meter, and they have precisely the same amount of mass; there are also two exactly identical meter sticks—their lengths are accurately at one meter, and they have precisely the same amount of mass (meter stick represents the tool of measuring length, width and height, the three dimensions that determine the size of space). And each meter stick is placed into the face of each of these two clocks, forming the shape like ⌀, thus becoming two sets of meter stick and clock, set A and set B. Then, one set of meter stick and clock, set A, is moving towards the massive celestial body under the action of its gravitational force, whereas another set of meter stick and clock, set B, still stays at the original location. In the eyes of the meter stick and clock of set B, the scales of space and time (being the length scale and time scale of set A) are gradually reduced as set A is moving towards this celestial body, because of the mass consumption from the mass of the meter stick and clock of set A, caused by their mass doing positive work (note: the mass consumption caused by mass doing positive work has been briefly introduced in the last section of the appendix of chapter six, P. 123 ~ 126).

One can easily simulate and experience the above description using the simple and convenient method as follows. First, fill air into two same balloons to the same size, say, to the diameter of one and a half meters, and mark them as balloon A and balloon B. Then, draw a meter stick of one-meter length and a clock of one-meter face on each of these two balloons. After that, pierce a small hole with a tiny needle on balloon A; you will see that the scales of the meter stick and clock on balloon A are gradually reduced (imitating the effect of the mass consumption from the mass of the meter stick and clock of set A described above), with respect to the scales on balloon B. That is, the scales of space and time on balloon A are gradually reduced, with respect to the scales on balloon B. (Narrator: once one has grasped this core concept, he or she has actually gotten the heart and soul of this new theory: it solves the fundamentally important problem of *why* and how space and time are variable thus relative in a gravitational field, by revealing and determining *why* and how the scales of space and time are variable thus relative in such a situation.)

This core concept has equipped us to go to one of the most important, also the most prominent, highlights of this new theory: *why* time runs slower and *why* length becomes shorter in a gravitational field (to the observers far away from it). Let us say, there are two observers, assigned as A and B; observer A (with clock A and meter stick A) is at a location near a massive celestial body (which is surrounded by the gravitational field generated by the celestial body from all directions), whereas observer B (with clock B and meter stick B) is at a position far away from the gravitational field (Fig. 4.1, along with the descriptions in the above two paragraphs). Now, if observer B reads the clock of observer A with the time scale on his own clock—clock B, he will find that time runs slower (that is, clock A runs slower than clock B), because his clock has a larger time scale than that of observer A. (Again, this is a bit like: when you read a clock of 10-centimeter face according to the time scale on a clock of 15-centimeter face, you will find that the clock of 10-centimeter face runs slower than the clock of 15-centimeter face. You can draw two such clocks on two transparent papers, and read them.)

Similarly, when observer B reads the meter stick of observer A with the length scale on his own meter stick—meter stick B, he will find that length becomes shorter (that is, meter stick A becomes shorter than meter stick B), because his meter stick has a larger length scale than that of observer A. (This is a bit like the following plain principle. Let us say, you measure the distance between two cities as 800 miles on one map, whose scale is 1:500; however, if you measure this distance on another map whose scale is 1:1000, but *still using the scale of 1:500*, you will measure this distance as only 400 miles.) Therefore, the core concept of this new theory clearly shows that this new theory reveals the mechanism thus the essence of: *why* and how time runs slower, *why* and how length becomes shorter in a gravitational field to the observers far away from it (because the scales of time and length are smaller in a gravitational field with respect to the scales in the region far away from the gravitational field).

All in all, the take home message about the core concept of this new theory is: the scales of space and time are reduced in a gravitational field, along with that: the stronger the gravitational field, the larger the reduction.

The principle of gravitational redshift in this new theory

Redshift, or shifted to the red end of the spectrum, is the increase in the wavelength of a ray of light or the decrease in its frequency during its traveling process. The redshift dealt with here is caused by a ray of light traveling along a direction in which the scales of length and time *increase*. For example, let us say that a ray of light passes through two different locations (location A and location B) that are far away from each other. The wavelength of the ray of light, when it passes through location A (where the length scale is 0.888), is measured as 660 nanometers (that is, a red light) by observer B at location B (where the length scale is 1.000; this measurement is with respect to his 1.000 length scale, of course). When the same ray of light travels to location B, its wavelength, still with respect to observer B and his 1.000 length scale, will become 743 nanometers (thus becoming redder); the value of 743 nanometers comes from 660 nanometers times (1.000/0.888) = 743 nanometers (note: one thousand million nanometers = one meter, or one million nanometers = one millimeter; millimeter is usually the smallest graduation on a ruler or a meter stick). In this example, redshift factor = (743 - 660)/660 = 0.126, being the increase rate of the wavelength of the ray of light, is equal to the increase rate of length scale, which is (1.000/0.888) - 1.000 = 0.126.

The principle of gravitational redshift in the newly discovered and verified gravitational theory (which is mechanism-revealed gravitational theory, MRGT, for short) is: because the scales of length and time *increase* radially outward (i.e., *decrease* radially inward) in the gravitational field of a massive celestial body (such as the sun or the earth), a ray of light that travels away from the celestial body is experiencing redshift (i.e., the wavelength of the ray of light becomes longer and longer or its frequency becomes lower and lower) (Fig. 4.2, next page). Therefore, this principle directly comes from the core concept of MRGT (this core concept has just been introduced in the section above). Because the redshift is caused by the existence of a gravitational field, it is thus referred to as gravitational redshift (which occurs when a ray of light travels along a direction in which the strength of a gravitational field *decreases*, thus the scales of length and time *increase*).

The magnitude of gravitational redshift is measured by the value of redshift factor or gravitational redshift factor; redshift factor is expressed with the increase rate of the wavelength of a ray of light or the decrease rate of its frequency. When expressed with the increase rate of the wavelength of a ray of light, redshift factor is equal to the increase rate of the length scale along the direction in which the ray of light travels. The key to understanding the principle of gravitational redshift is: the scales of length and time *increase* radially outward (i.e., *decrease* radially inward) in the gravitational field of a massive celestial body; that is, the key to understanding this principle lies with firmly grasping the core concept of MRGT you have seen above.

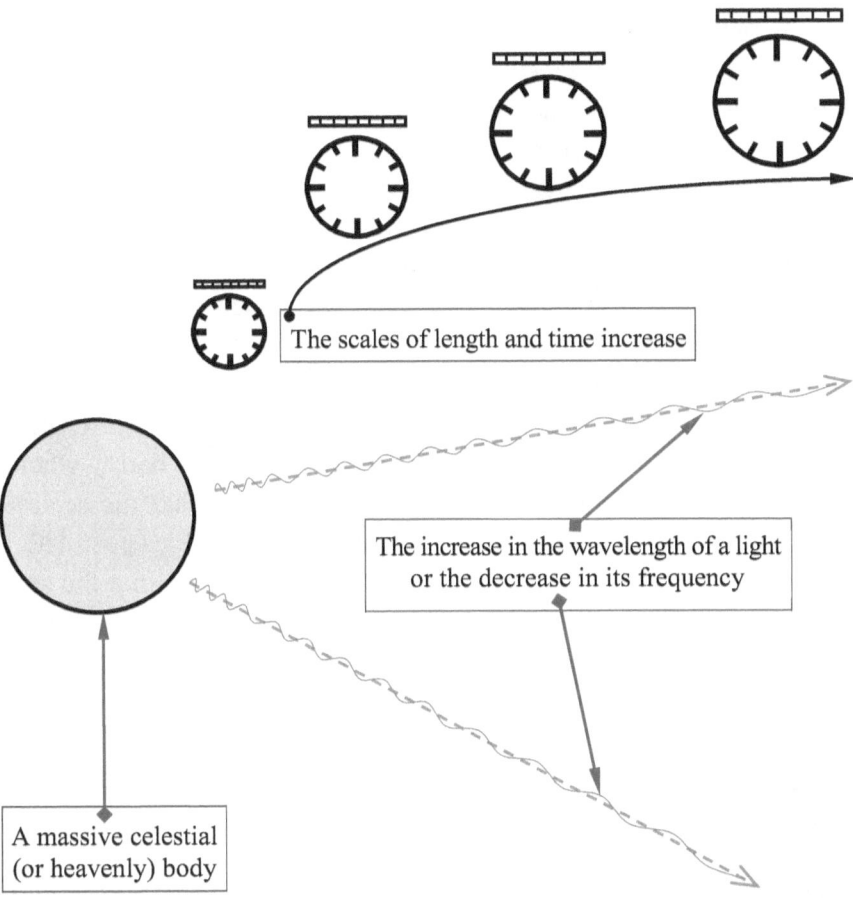

Figure 4.2, an illustration of the principle of gravitational redshift in the newly discovered and verified gravitational theory (which is mechanism-revealed gravitational theory); this principle reveals the mechanism and essence of gravitational redshift.

The principle of gravitational redshift (in the newly discovered and verified gravitational theory, which is mechanism-revealed gravitational theory) reveals the mechanism of gravitational redshift as that: when a ray of light travels away from a gravitational field, its wavelength is continuously becoming longer and longer, or its frequency is continuously becoming lower and lower, because the scales of length and time continuously *increase* along the direction in which the ray of light travels. Therefore, the essence of gravitational redshift is that a ray of light travels away from a gravitational field along a direction in which the scales of length and time *increase*, because the strength of the gravitational field *decreases* in this direction. (Related questions and answers: can the concept of gravitational redshift interpreted with general relativity reveal the mechanism and essence of gravitational redshift? If not, why? Answer: it cannot. This is because the mechanism and essence of gravitational redshift is that a ray of light travels along a direction in which the scales of length and time *increase* in the gravitational field generated by a massive celestial body, whereas general relativity literally **assumes** or **hypothesizes** that the scales of length and time are *invariable* via its indispensable postulate of 'invariant scales of length and time', which simply says that the scales of length and time at different points over an entire gravitational field are the same. And so, general relativity is unable to reveal the mechanism and essence of gravitational redshift. Question: what is the fundamental reason of this inability? Answer: because general relativity is unable to solve the most fundamental problem in front of itself, which is *why* space and time are variable thus relative in a gravitational field, or *why* time runs slower and *why* length becomes shorter in such a situation. Question: what is the hard evidence of this inability? Answer: this hard evidence is the plain *truth* that general relativity indispensably and desperately necessitates its indispensable postulate of 'invariant scales of length and time', simply because this plain *truth* turns out to be the sufficient and explicit, also undeniable or irrefutable, evidence that general relativity is really unable to solve this most fundamental problem, as analyzed and unveiled in chapter three.)

What should be mentioned is: the principle of gravitational redshift (in the newly discovered and verified gravitational theory, which is mechanism-revealed gravitational theory, or MRGT, for short) has been verified by citing the two famous, also highly authoritative, experimental or observational results—both results have been fully recognized by the scientific community. (i) By citing the result in the famous Harvard tower experiment that was done in 1959 (the purpose of this experiment was to verify the gravitational redshift very near the surface of the earth—i.e., immediately above the surface of the earth. This famous experiment, which has been widely regarded as a crucial confirmation of the gravitational redshift interpreted in general relativity, is also referred to as Pound-Rebka experiment nowadays). The calculated value (with the equation of the principle of gravitational redshift; this equation is a very important equation in MRGT) is 2.442×10^{-15}; the widely accepted value is 2.455×10^{-15}. Therefore, this calculated value accurately agrees with the widely accepted value. And so, it is rather rational that we ought to be highly confident in the principle of gravitational redshift and MRGT, even though they are quite new. (ii) By citing the result obtained from the measurement of the gravitational redshift of the sun (this measurement was carried out for the first time in 1962 by a team from Princeton University). The calculated value (with the equation of the principle of gravitational redshift in MRGT) is 1.225×10^{-6}; the widely accepted value is 1.23×10^{-6}. Therefore, this calculated value perfectly agrees with the widely accepted value. And so, it is quite rational that we ought to be fully, at least highly, confident in the principle of gravitational redshift and MRGT, albeit they are new.

What should be pointed out is that the attribute of the principle of gravitational redshift is of mechanism-revealed nature, because this principle is simply a direct application of MRGT, whose attribute is of mechanism-revealed nature, as mentioned above; because in this application, none of the assumptions, hypotheses and postulates (AHPs, for short) has been employed; in fact, none of the AHPs is necessary at all in this application. (Friendly reminder: what should be noticed is that the attribute of the principle of gravitational redshift explained with MRGT is *fundamentally different* from that of the concept of gravitational redshift interpreted with general relativity, whose attribute is of postulate-based nature, as mentioned above; as a result—as a clear and definite result, also an inevitable and unavoidable result, the attribute of this concept is of postulate-based nature.)

The principle of gravitational light bending in this new theory

The principle of gravitational light bending in the newly discovered and verified gravitational theory (which is mechanism-revealed gravitational theory, MRGT, for short) reveals the mechanism and essence of why and how a ray of light is bent when it passes through a gravitational field. This principle is the direct result of the combination of the two things: one is the core concept of MRGT, which is the gravitational scale contour lines of space and time in the gravitational field of a massive celestial body (the sun or the earth, for example); another is the inertia principle of photons traveling (photons are the tiny and discrete particles of light, so a photon can be regarded as a ray of light from the angle of comprehension).

The inertia principle of photons traveling refers to the nature of photons tending to travel along the direction of equal space scale or equal time scale; that is, a ray of light tends to travel along the direction of equal space scale or equal time scale in a gravitational field. Because the gravitational scale contour lines of space and time in the gravitational field of a massive celestial body (the sun or the earth, for example) consist of an infinite number of concentric spheres, with the mass center of the body as their common center, the inertia principle of photons traveling determines that a ray of light tangentially passes one point on a gravitational scale contour line of space and time at an earlier moment, the ray of light also tangentially passes another different point on the same gravitational scale contour line at a later moment, a bending angle is thus formed by the traveling route of the ray of light in the gravitational field (e.g., a bending angle is formed by a ray of light's traveling route along the path of A → B → C → D → E, Fig. 4.3, next page; note: a gravitational scale contour line of space and time is a circle in 2-D, a sphere in 3-D).

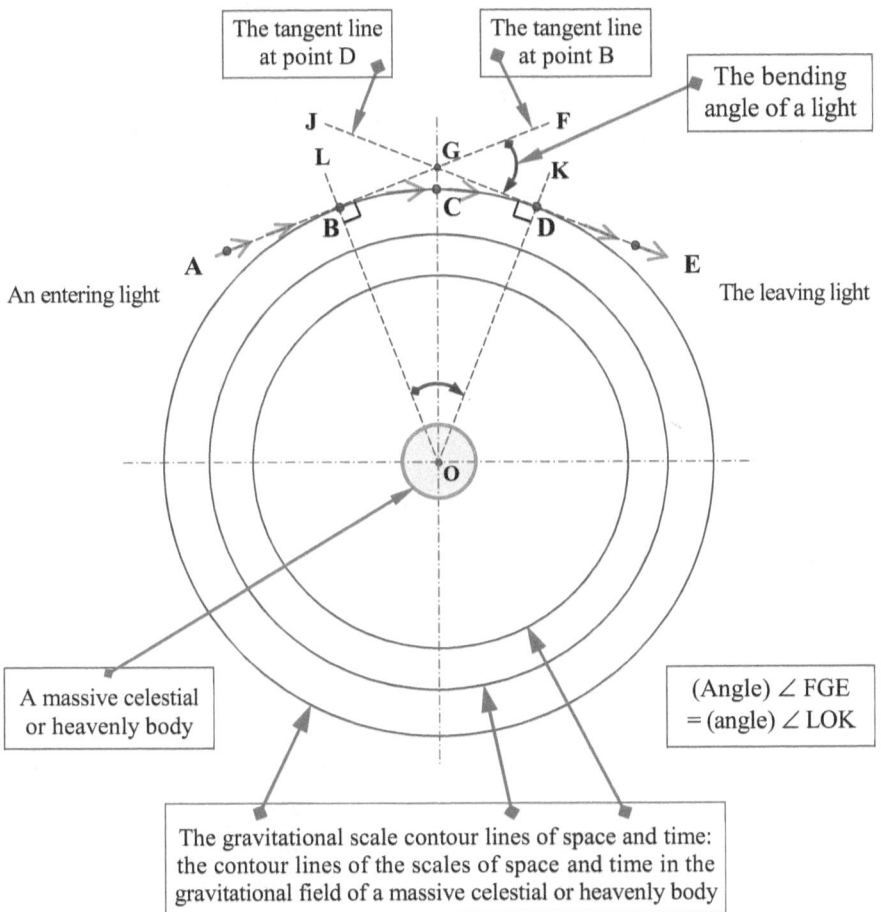

Figure 4.3, an illustration of the principle of gravitational light bending in the newly discovered and verified gravitational theory (which is mechanism-revealed gravitational theory); this principle reveals the mechanism and essence of why and how a ray of light is bent when it passes through the gravitational field of a massive celestial body (a star with a large amount of mass, e.g.).

The gravitational light bending equation (GLBE) (in the newly discovered and verified gravitational theory, which is mechanism-revealed gravitational theory, or MRGT, for short) is the mathematical expression of the principle of gravitational light bending just mentioned above. GLBE determines the relationship between the bending angle of light and the reduction in length scale, or the relationship between the bending angle of light and the reduction in time scale (note: the reduction in length scale is equal to the reduction in time scale, as revealed in MRGT). GLBE, by showing that the bending angle of light is the function of the reduction in length scale or the function of the reduction in time scale, reveals that the stronger a gravitational field, the larger the reduction in length scale and time scale in the gravitational field; thus the larger the bending angle of light in the gravitational field. For instance, according to the results calculated with GLBE, the bending angle of light on the location of the average orbit of the earth due to the gravitational field of the sun is only 1/215 of the bending angle of light closely around the sun; the bending angle of light on the surface of the earth due to the gravitational field of the earth itself is only 1/3045 of the bending angle of light closely around the sun.

When viewed and illustrated geometrically, the value of the bending angle of a light, which is equal to the angle formed by the two tangent lines at the two points (e.g., points B and D in Fig. 4.3) on the same gravitational scale contour line that is tangentially passed by the light, is equal to the central angle swept by the traveling route of the light on the circle consisting of this gravitational scale contour line, that is, $\angle$ FGE = $\angle$ LOK (Fig. 4.3). (Note: there is such a simple relationship in primary geometry: the angle formed by the two tangent lines of a circle is equal to the central angle of the same circle.)

The take home message about the principle of gravitational light bending is: the inertia principle of photons traveling makes a light tend to travel along the direction of equal space scale or equal time scale in a gravitational field. Or stated concisely and plainly, the mass of a massive celestial body (the sun or the earth, for example) marks the scales of space and time, the inertia principle of photons traveling makes a ray of light travel along the direction of the marked scales of space and time.

What should be mentioned is that the value of the bending angle of light (at the location of closely around the sun) calculated with the gravitational light bending equation in MRGT (which is 1.750 arcsec) accurately agrees with the observed or measured value (which is 1.75 arcsec) (note: arcsec is a very small unit of angle, much smaller than degree—that is, one arcsec is much smaller than one degree. Then how large is one degree? Please notice that a right angle is equal to 90 degrees). This accurate agreement thus shows that the principle of gravitational light bending in MRGT has been verified via this famous, observed value. (Commentator: with this verification, it becomes rationally reasonable that we ought to be confident in the principle of gravitational light bending and MRGT, even though they are quite new.)

*What should be pointed out is: while the result calculated with general relativity also accurately agrees with this famous, observed value, the calculation with the gravitational light bending equation in MRGT is much, much simpler than the calculation with general relativity. Why? The fundamental reason is that the attribute of MRGT is of mechanism-revealed nature, whereas the attribute of general relativity is of postulate-based nature (the different attributes of these two theories have been explicitly mentioned earlier in this chapter). And so, this instance opens a window through which one could clearly see such a fact: what general relativity can do, MRGT can do also; but MRGT is much, much simpler. Of course, what's far more important is that MRGT can do what general relativity cannot. For example, MRGT has solved the fundamental problem of *why* space and time are variable thus relative in a gravitational field (by revealing and determining *why* and how the scales of space and time are reduced in such a situation), or has shown *why* time runs slower and *why* length becomes shorter in a gravitational field (by revealing the mechanism behind these two *whys*), as presented above; whereas general relativity doesn't and can't solve this fundamental problem, as analyzed and unveiled in chapter three. Why? Again, the root cause is that the attribute of MRGT is of mechanism-revealed nature, whereas the attribute of general relativity is of postulate-based nature.

Chapter 5

A Fundamentally New Black Hole Theory: Mechanism-Revealed Black Hole Theory

[The Window on This Chapter]

Mechanism-revealed black holes are the black holes revealed and explained by a new black hole theory, which is mechanism-revealed black hole theory that is based on the new gravitational theory (mechanism-revealed gravitational theory) introduced in chapter four.

A mechanism-revealed black hole is a hugely massive celestial body that reduces the scales of length and time in its vicinity to such an extent that all visible light becomes invisible—all the visible light entering the region of the black hole becomes invisible; all the light emitted from the black hole is also invisible in the region of the black hole, resulting in this region looking totally black to the observers far away from it.

The gravitational nature of mechanism-revealed black holes is fundamentally and essentially the same as other ordinary celestial bodies, even though the gravitational effects caused by mechanism-revealed black holes, due to their hugely massive feature, are much, much stronger and greater than other ordinary celestial bodies.

Only mechanism-revealed black holes can reveal the mechanism and essence of black holes, and can reveal the constituents of black holes, thus can uncover the fundamental nature of black holes; that is to say, only mechanism-revealed black holes can solve the fundamental, essential and crucial problems about black holes.

Only mechanism-revealed black holes can unveil the long-standing mystery of dark matter (this mystery includes the constituents and fundamental nature of dark matter).

The two basic features of this new black hole theory

This new black hole theory (which was developed not very long ago by me based on the new gravitational theory introduced in chapter four) has the following two basic features. First, as its fundamental and easily noticeable mark, the attribute of this new black hole theory is of mechanism-revealed nature, being clearly determined by two sufficient and explicit reasons. One reason is that its theoretical basis is of mechanism-revealed nature. The theoretical basis of this new black hole theory is the newly discovered and verified gravitational theory (which is mechanism-revealed gravitational theory, or MRGT for short; the attribute of MRGT is of mechanism-revealed nature, as explicitly pointed out in chapter four) that has revealed the mechanism of *why* space and time are variable thus relative in a gravitational field, or that has revealed the mechanism of *why* time runs slower and *why* length becomes shorter in a gravitational field. Another reason is that this new black hole theory has revealed the mechanism behind its describing phenomena (this will be presented in the next section). For these two reasons, this new black hole theory has been named as *mechanism-revealed black hole theory* (MRBHT, for short) by me. Accordingly, black holes, when they are explained with MRBHT, are referred to as *mechanism-revealed black holes*.

*What should be pointed out or realized here is: the third important reason that I have named this new black hole theory as *mechanism-revealed* black hole theory (MRBHT, for short) is to distinguish MRBHT from all other black hole theories that are based on general relativity, whose attribute is of *postulate-based* nature. (Related knowledge or friendly reminder: as mentioned in chapter four, general relativity is based on at least five assumptions, hypotheses and postulates, thus the real or accurate name of general relativity is (or ought to be) ***postulate-based*** **general relativity**, because its attribute is obviously of *postulate-based* nature. Therefore, all the black hole theories based on *postulate-based* general relativity are *postulate-based* black hole theories. Accordingly, also unavoidably, black holes, when they are interpreted with the black hole theories based on *postulate-based* general relativity, are inevitably *postulate-based black holes*, either in essence or in truth or in both. Moreover, please be reminded more specifically thus more explicitly: the black hole theories established by Stephen Hawking, because they are based on *postulate-based* general relativity, are *postulate-based* black hole theories; and so, black holes, when they are interpreted with Hawking's black hole theories, are *postulate-based black holes*. In fact, virtually all the existing black hole theories, except the newly developed MRBHT, are based on *postulate-based* general relativity. For instance, the related work on black holes done by the German astronomer Karl Schwarzschild in the early 20th century was based on the theory of general relativity soon after this theory had been developed. The related result about black holes obtained by Robert Oppenheimer is also based on general relativity. The basic feature of non-rotating black holes discovered by Werner Israel is based on general relativity too. The discovery of the features of rotating black holes by Roy Kerr is based on general relativity again.) (Friendly and related reminder: as analyzed and unveiled in chapter three, general relativity, or *postulate-based* general relativity, turns out to be a theory that doesn't and can't have the ability to solve the most fundamental problem in front of itself, which is *why* space and time are variable thus relative in a gravitational field, or *why* time runs slower and *why* length becomes shorter in such a situation.)

Second, the characteristic, also noticeable and impressive, feature of this new black hole theory (that is, MRBHT stands for mechanism-revealed black hole theory) lies with: it is easy to understand. For instance, the theoretical core of MRBHT is similar in essence to the following plain principle. Let us say, you measure the distance between two locations as 760 miles on one map, whose scale is 1:250. If you measure this distance on another map whose scale is 1:500, but *still using the scale of 1:250*, you will measure this distance as 380 miles. Why is MRBHT so simple? The fundamental reason is that MRBHT is of mechanism-revealed nature, as just pointed out above; the direct reason is that MRBHT, because it is of mechanism-revealed nature, is mathematically pretty simple: the mathematics at the level of high school or first-year undergraduate is enough.

The mechanism and essence of this new black hole theory

Having previewed the two basic features of this new black hole theory, we are about to enter its mechanism and essence—the specific and key information of this theory. Whenever a black hole is talked about, light is always an indispensable, also crucially important, subject. So, as a preparation to enter the mechanism and essence of this new theory, let us review the basic concept about visible and invisible light first.

This concept includes two aspects. First aspect, whether a light is visible or invisible depends on its wavelength or frequency (the wavelength of a light is the distance between its one wave crest and the next; the frequency of a light is the number of its waves per second). For example, it has been known that the wavelength of visible light is in the range from 380 to 760 (or 390 to 780) nanometers, with violet light between 380 to 440 nanometers, and red light between 620 to 760 nanometers (note: one thousand million nanometers = one meter, or one million nanometers = one millimeter; millimeter is usually the smallest graduation on a ruler or a meter stick). Second aspect, the wavelength of a light is determined by the length scale used to measure its wavelength (correspondingly, the frequency of a light is determined by the time scale used to measure its period or frequency; its period times its frequency is equal to one). For instance, the wavelength of a light is measured as 700 nanometers (that is, a red light) by observer A at location A where the local length scale is 1.00. When the same light travels to location B where the local length scale is 0.60, what is the wavelength of the same light with respect to observer A (that is, still with respect to his 1.00 length scale)? Answer: the wavelength of the same light is equal to 420 nanometers, which is from 700 nanometers times (0.60/1.00), thus becoming a violet light.

Now, let us go to the *mechanism* and *essence* of this new black hole theory (that is, mechanism-revealed black hole theory). According to the newly discovered and verified gravitational theory (which is mechanism-revealed gravitational theory, some of its main and key results have been briefly introduced in chapter four), the length scale and time scale in the gravitational field of a massive celestial body increase radially outward (decrease radially inward) (Fig. 5.1, next page). That is, the farther away from the body, the larger the scales of length and time; the closer to the body, the smaller the scales of length and time. As a result, when the mass of a **hugely** massive celestial body reduces the scales of length and time in its vicinity to such an extent that all visible light becomes invisible, a black hole is thereby formed. For example, the wavelength of a light is measured as 700 nanometers (that is, a red light) by observer George at a location very far away from a **hugely** massive body, where the length scale is infinitely close to 1.00 (thus can be practically regarded as 1.00). When the same light travels to a location very near the **hugely** massive body, where the length scale is 0.50, to observer George (of course, still with respect to his 1.00 length scale), the wavelength of the same light will become 350 nanometers (thus becoming an invisible light)—this 350 nanometers comes from: 700 nanometers times (0.50/1.00).

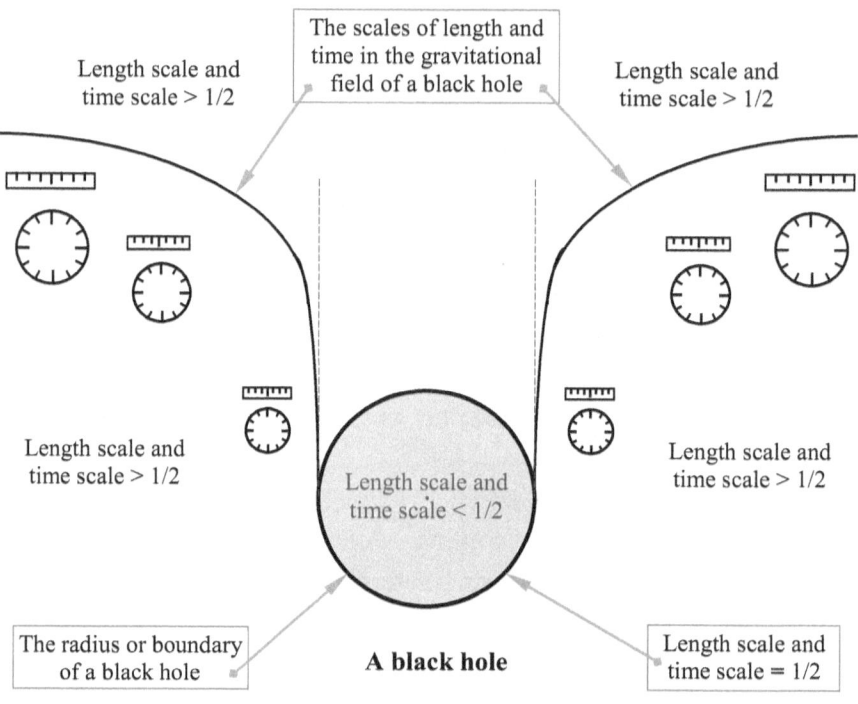

Figure 5.1, mechanism-revealed black hole theory (a new black hole theory). This new black hole theory reveals the mechanism and essence of black holes, thus solves the fundamental, essential and crucial problems about black holes (that is, all the long-term unsolved, greatly important problems about black holes have been solved by this new black hole theory).

So a black hole is a region closely around a **hugely** massive celestial body, where all visible light becomes invisible: all the visible light entering the region of the black hole becomes invisible; all the light emitted from the black hole is also invisible in the region of the black hole. Resultantly, to very distant observers, there are no visible light rays at all in the region of the black hole—this region looks totally black. Therefore, the *mechanism* of mechanism-revealed black hole theory is that the existence of a **hugely** massive celestial body reduces the scales of length and time in its vicinity to such an extent that all visible light becomes invisible; thus the *essence* of a (mechanism-revealed) black hole is the tremendous reduction in the scales of length and time that occurs in the gravitational field of (and near) a **hugely** massive celestial body. Moreover, what should be mentioned or noticed is: the length scale and time scale in the enormously large extending region that radially spreads out from a black hole are significantly reduced too (due to the existence of the black hole), though not as intense and obvious as within the region of the black hole.

After seeing the *mechanism* and *essence* of mechanism-revealed black hole theory, let us go to the definition of a mechanism-revealed black hole, which combines and integrates this *mechanism* and *essence*. This definition is: **a mechanism-revealed black hole is a hugely massive celestial body** that reduces the scales of length and time in its vicinity to such an extent that all visible light becomes invisible—all the visible light entering the region of the black hole becomes invisible; all the light emitted from the black hole is also invisible in the region of the black hole, resulting in this region looking totally black to the observers far away from it (so the most fundamental core point of this definition is: **a mechanism-revealed black hole is a hugely massive celestial body**). This definition shows and determines that the gravitational nature of mechanism-revealed black holes is fundamentally and essentially the same as other ordinary celestial bodies, even though the gravitational effects caused by mechanism-revealed black holes, due to their **hugely** massive feature, are much, much stronger and greater than other ordinary celestial bodies.

The size of a mechanism-revealed black hole

After knowing about the mechanism and essence of mechanism-revealed black hole theory (MRBHT) as well as the definition of a mechanism-revealed black hole based on this mechanism and essence, let us see the size of a mechanism-revealed black hole, determined with MRBHT.

In order to calculate the size of a mechanism-revealed black hole, we need to know the meaning of the ratio of <u>v</u>isible <u>l</u>ight <u>b</u>oundary wavelength (assigned as W_{vlb}). The idea of W_{vlb} is from color index, the ratio of the wavelength of violet to red light. As mentioned above, the wavelength of visible light is in the range from 380 to 760 (or 390 to 780) nanometers (note: one thousand million nanometers = one meter), which means that the lower and upper boundaries of visible light's wavelength are respectively 380 and 760 nanometers (or 390 and 780 nanometers), with violet light close to the lower boundary, and red light close to the upper boundary. Thus W_{vlb} is equal to 380/760 (or 390/780) = 1/2. So, W_{vlb} is the result of an application of color index, by extending color index to the very edge of the wavelength of visible light, in order to reflect the black feature of black holes.

Now, we are ready to see the size of a mechanism-revealed black hole, represented with its radius (assigned as R_{bh}). The radius of a mechanism-revealed black hole is the threshold radius that marks the black hole's boundary, on which entering visible light just begins to become invisible, whereas leaving light just begins to become visible. This determines that the radius of a mechanism-revealed black hole must combine the mechanism of mechanism-revealed black hole theory with the meaning of W_{vlb} (this meaning has been clearly elucidated in the paragraph above), in order to provide a reasonable thus easily comprehensible explanation for the observed phenomenon of black holes.

This combination determines that the value of radius R_{bh} corresponds to the value of r in the quantity $GM/(rc^2)$ that makes length scale = W_{vlb} = 1/2 (where M is the mass of a **hugely** massive celestial body that results in a black hole, r is the radius from the mass center of the mass M, G is the universal gravitational constant, and c is the speed of light. So the quantity $GM/(rc^2)$ is unitless. *Oh, by the way, if the r in the quantity $GM/(rc^2)$ is equal to R, where R is the radius of a celestial body, this quantity thus becomes $GM/(Rc^2)$. What should be mentioned is that $GM/(Rc^2)$ is one of the most important quantities in modern astronomy; it is no exaggeration to say that all today's astronomers are very familiar with this specific quantity. As a result, such an important quantity in astronomy, $GM/(Rc^2)$, turns out to be a special case of the quantity $GM/(rc^2)$; this feature can enable one to perceive and realize the meaning and importance of the quantity $GM/(rc^2)$ more clearly and more impressively[*]). $GM/(rc^2)$ is a crucial term in the core equation of mechanism-revealed gravitational theory (this theory has been briefly introduced in chapter four; and its core equation, which is the equation of gravitational scales of space and time, determines the length scale and time scale in the gravitational field of a massive celestial body). Then, by letting the length scale in this core equation be equal to W_{vlb} = 1/2, obtain $GM/(rc^2) = 2$. Finally, from $GM/(rc^2) = 2$, obtain $r = R_{bh} = GM/(2c^2)$, where M is the mass of a black hole, G and c have the same meaning as above. (For comparison, the value of this radius is exactly one-fourth of that of the well-known Schwarzschild radius.)

Now, one can clearly see that according to mechanism-revealed black hole theory, the size of a mechanism-revealed black hole, when expressed by its radius, is proportional to its mass; that is to say, the size of a mechanism-revealed black hole entirely and only depends on its mass, or the size of a mechanism-revealed black hole is totally determined by its mass, being an explicit, also remarkably important, basic feature of mechanism-revealed black holes. (Interestingly, this basic feature is clearly similar to the conclusion that the size of a black hole depends only on its mass, an important conclusion obtained from the available black hole theories based on general relativity; for instance, this conclusion has appeared in Hawking's *A Brief History of Time*. And this basic feature is also quite similar to the result that the area of a black hole increases whenever matter falls into it, an important result gotten from the available black hole theories based on general relativity; for instance, this result has also been mentioned in Hawking's *A Brief History of Time*. Quite obviously, these similarities can substantially enhance the acceptability of this new black hole theory, at least to a certain degree or from a certain angle.)

Despite the similarities above, what should be pointed out is that the radius of a mechanism-revealed black hole in mechanism-revealed black hole theory is fundamentally different from the event horizon of a postulate-based black hole (the boundary of a postulate-based black hole) in those postulate-based black hole theories based on postulate-based general relativity. This fundamental difference is: the radius of a mechanism-revealed black hole is the dividing line between visible and invisible light; whereas the event horizon of a postulate-based black hole is defined as the boundary from which light rays just fail to escape from the black hole (the event horizon of a postulate-based black hole is a crucially important concept in the postulate-based black hole theories proposed by Stephen Hawking).

The three fundamental natures of mechanism-revealed black holes

Mechanism-revealed black hole theory (MRBHT) reveals three fundamentally important natures of mechanism-revealed black holes. Because these natures are of fundamental, essential and crucial importance, let us go over them one by one in this section, specifically and concisely.

The first nature is about the constituents of mechanism-revealed black holes. In comparison with other ordinary celestial bodies (such as stars and planets), the constituents of mechanism-revealed black holes are not fundamentally different at all—neither mysterious nor unique in essence; that is to say, fundamentally speaking, the constituents of mechanism-revealed black holes are the same as other ordinary celestial bodies (thus, the gravitational nature of mechanism-revealed black holes is totally the same as other ordinary celestial bodies, even though the gravitational effects caused by mechanism-revealed black holes, due to their hugely massive feature, are much, much stronger and greater than other ordinary celestial bodies). Please notice that, as revealed by MRBHT, the decisive difference between (mechanism-revealed) black holes and other ordinary celestial bodies (such as a variety of stars and planets) lies with the **hugely massive feature** of black holes, rather than in their constituents. In other words, a (mechanism-revealed) black hole can be formed as a result of many ordinary celestial bodies joining together. For instance, according to MRBHT, as long as the mass of a celestial body is greater or equal to several dozens of times the mass of the sun, the celestial body is or becomes a (mechanism-revealed) black hole, if the average density of the black hole is set at the density of an atomic nucleus. The available knowledge about the mass of various celestial bodies shows that such a requirement on mass is not difficult to meet even in (our) the Milky Way galaxy, needless to say in the vast universe. (In contrast, according to *postulate-based black holes*, being the black holes interpreted with those *postulate-based* black hole theories based on *postulate-based* general relativity, the constituents of black holes have been, and are, a great mystery; in fact, within the paradigm or stereotype of *postulate-based black holes*, the constituents of black holes have actually become one of the long-term unsolved, fundamentally important problems in science or in physics—in modern astronomy or in astrophysics, to be exact. Note: astrophysics is a branch of astronomy that deals with the physical and chemical structure of the stars, planets, etc.)

*Related clarification for avoiding possible confusion. The above conclusion, which is that the constituents of (mechanism-revealed) black holes are the same as other ordinary celestial bodies, neither excludes the factor or possibility that different (mechanism-revealed) black holes may have very different constituents, nor denies that the constituents of different types of ordinary celestial bodies (such as different types of stars) are different greatly. In addition, this conclusion does not exclude the factor or possibility that the constituents of (mechanism-revealed) black holes can be much more composite and much more inclusive than other ordinary celestial bodies, because a (mechanism-revealed) black hole is much, much more massive than an ordinary celestial body.

Second nature: a mechanism-revealed black hole can emit light, even a very large amount of light rays with great intensity and high density, from the region within its radius, though the emitted light is invisible in the region within the radius of the black hole (note: light rays are also called photons). This is because a (mechanism-revealed) black hole, as just mentioned above, is a region closely around a hugely massive celestial body, where all visible light becomes invisible: all the visible light entering the region of the black hole becomes invisible; all the light emitted from the black hole is also invisible in the region of the black hole. And so, to very distant observers, there are no visible light rays at all in the region of the black hole—this region looks totally black. This nature enables mechanism-revealed black hole theory to be the key to solving the four long-standing, fundamentally important problems in science: dark matter, gamma ray bursts, ultrahigh-energy cosmic rays, and the GZK paradox (this paradox has been concisely explained in the Glossary of this book). (On the contrary, according to the *postulate-based* black hole theories based on *postulate-based* general relativity, a black hole cannot emit light from the region within its boundary, the so-called event horizon. Consequently, these *postulate-based* black hole theories are not only unable to solve these four big problems, but have also become the shackles and obstacles to solving them.)

*What should be emphasized or clarified is that, though Hawking has mentioned the emission or radiation from (postulate-based) black holes (in his *A Brief History of Time*, for example)—that is, a (postulate-based) black hole can emit X rays and gamma rays, this emission is merely from the place just outside the event horizon of the black hole, rather than from within the black hole. Please further notice that, the existence of such a sort of emission is under the fundamental *prerequisite* that **nothing** can escape from within the event horizon of a (postulate-based) black hole. Moreover, this sort of emission is far too small to explain the **tremendously huge** sources of gamma ray bursts (otherwise, the big puzzle of gamma ray bursts would have already been solved much earlier!). Therefore, there is an essential difference between the emission from *postulate-based* black holes and the light rays (also called photons) emitted from *mechanism-revealed* black holes. (Related question and answer: what is the fundamental and direct reason causing this essential difference? Answer: within the paradigm or stereotype of *postulate-based* black holes, the definition of the event horizon is: the event horizon of a black hole is formed by the light rays that just fail to escape from the black hole; such a definition clearly tells us that a black hole cannot emit light from the region within its boundary, the so-called event horizon. However, the event horizon of a black hole (the boundary of a black hole) turns out to be an unreliable concept for two reasons. One is that the concept of the event horizon originates from the *momentum* of photons, but the current method of calculating the *momentum* of photons, which is the very method that determines the value of the *momentum* of photons, turns out to be clearly wrong, because it has led to obviously ridiculous, plainly unreasonable results. Another reason is that the concept of the event horizon turns out to be self-contradictory. For the detailed analysis and specific scrutiny of *why* the event horizon of a black hole (the boundary of a black hole) turns out to be an unreliable concept due to these two reasons, please refer to the appendix of this chapter, P. 86 ~ 94.)

In addition, a mechanism-revealed black hole does not have the so-called singularity; in fact, a mechanism-revealed black hole does not need the concept of singularity at all. (Related knowledge: as mentioned in the last section of chapter three, the so-called singularity, being predicted or created by *postulate-based* general relativity, is a mathematical point whose size is zero or infinitely close to zero and *with infinite density and infinite temperature*. However, believe it or not, physically the two aspects of the concept of singularity, which are '*infinite density and infinite temperature*', are actually incompatible thus clearly self-contradictory in essence. Specifically, please carefully notice that these two aspects cannot coexist at all: infinite density denies motion, whereas infinite temperature requires a motion at a very near the speed of light. As a result—as an explicit and undeniable result in truth, the concept of singularity turns out to be clearly and seriously self-contradictory, being an obvious and undeniable feature of singularity. Another prominent, also irrefutable or undeniable, feature of singularity is that general relativity itself breaks down at singularity, being a fully recognized feature of singularity by the scientific community in physics. In other words, the concept of singularity actually and unavoidably tells us such a dilemma: general relativity has created a place (singularity) where general relativity itself is no longer workable or applicable.) Therefore, at any location or place within or around a mechanism-revealed black hole, there is utterly no self-contradictory point at all.

(In contrast, according to the *postulate-based* black hole theories based on *postulate-based* general relativity, there is a singularity inside a black hole. However, the concept of singularity turns out to be clearly and seriously self-contradictory, as just analyzed and pointed out above.) (Related question and answer: what is the fundamental reason or cause that *postulate-based* general relativity creates or predicts the so-called singularity? Answer: *postulate-based* general relativity does not have the ability to solve the most fundamental problem in front of itself, which is *why* space and time are variable thus relative in a gravitational field, or *why* time runs slower and *why* length becomes shorter in such a situation, as unveiled and pointed out at the beginning of the last section of chapter three.)

The great gravitational redshift caused by mechanism-revealed black holes

The gravitational redshift caused by mechanism-revealed black holes is explained by the principle of gravitational redshift (in the newly discovered and verified gravitational theory, which is mechanism-revealed gravitational theory, MRGT, for short; this principle has been concisely introduced in chapter four), for two sufficient reasons. First, the gravitational nature of mechanism-revealed black holes is the same as other ordinary celestial bodies, because the constituents of mechanism-revealed black holes are the same as other ordinary celestial bodies, as revealed by mechanism-revealed black hole theory (MRBHT, for short) and as mentioned in the section above. Second, the principle of gravitational redshift reveals the mechanism and essence of gravitational redshift, as pointed out in chapter four (for the specific information about the mechanism and essence of gravitational redshift, please go back to the related section of chapter four if necessary).

In addition to the two sufficient reasons above, there are two facts in the relationship between the principle of gravitational redshift and MRBHT. One fact is that this principle directly comes from the core concept of MRGT; that is, the core concept of MRGT is the direct theoretical basis of this principle. Another fact is that the core concept of MRGT is also the direct theoretical basis of MRBHT. Thus, both the principle of gravitational redshift and MRBHT are attached onto the same thing—the core concept of MRGT. Therefore, these two facts are also a further substantiation or confirmation that the gravitational redshift caused by mechanism-revealed black holes ought to be explained by the principle of gravitational redshift. Specifically, this principle is illustrated via the gravitational redshift caused by a mechanism-revealed black hole (Fig. 5.2, next page).

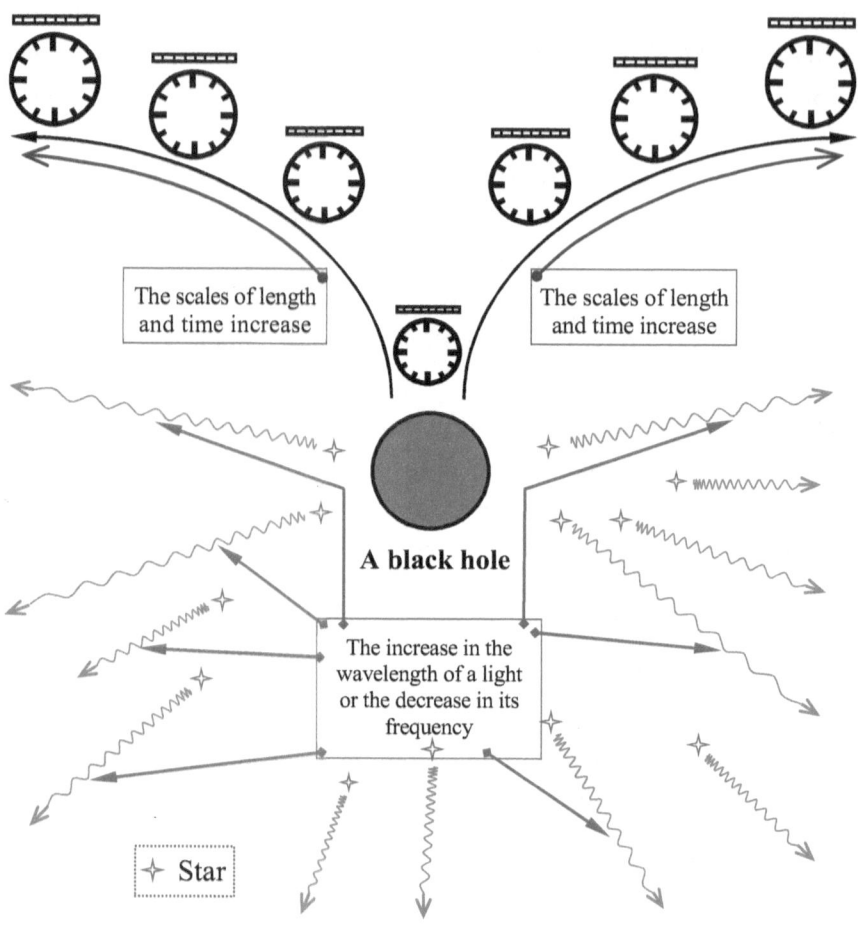

Figure 5.2, the gravitational redshift caused by a (mechanism-revealed) black hole.

The magnitude of gravitational redshift is measured by the value of gravitational redshift factor; this value is calculated with the equation of the principle of gravitational redshift, one of the most important equations in MRGT. Gravitational redshift factor is expressed with the increase rate of the wavelength of a ray of light or the decrease rate of its frequency. When expressed with the increase rate of the wavelength of a ray of light, the value of gravitational redshift factor is equal to the increase rate of the length scale along the direction in which the ray of light travels.

Due to their hugely massive feature, mechanism-revealed black holes can cause **very large** gravitational redshift. That is to say, due to the existence of a mechanism-revealed black hole, there can be **very large** gravitational redshift that occurs in the very big extending region radially spreading out from the black hole. For instance, at the distances of 2, 4, 8, 16, 32, 40, 50 times of the radius of a mechanism-revealed black hole from the center of the black hole, the value of gravitational redshift factor, when expressed with the increase rate of the wavelength of a ray of light, is respectively equal to 0.6180, 0.3660, 0.2071, 0.1124, 0.0590, 0.0477, and 0.0385. This means that the increase rate of the wavelength of light is respectively equal to 61.80%, 36.60%, 20.71%, 11.24%, 5.90%, 4.77%, and 3.85%, if light rays (from the stars in the great extending region that radially spreads out from the black hole) are emitted radially away from the black hole from the positions of these distances (Fig. 5.2).

What should be pointed out here is that the size (or the radius) of a black hole can be very large: the size of a big black hole, whose mass can be hundreds even thousands of times the mass of the sun—this is not unusual from the present knowledge about black holes, can be quite comparable to the size of the sun. Moreover, it has been known that the existence of a black hole, due to its highly massive feature (the mass of a typical black hole can be hundreds of times that of the sun), can cause the gravitational field in the vast extending region outside it to be significantly stronger, actually much stronger, than without the black hole. (Related knowledge: the region significantly affected by a black hole is much, much larger than that occupied by the black hole, often millions, even billions, of times larger, because there is a hugely vast extending region radially spreading out from a black hole.)

More specifically, also more impressively, it has been recognized that there is a supermassive black hole at the center of the Milky Way galaxy; in fact, at the center of each observable galaxy, there is also a supermassive black hole. Furthermore, the mass of each supermassive black hole of this type is in the range of millions, even billions, of times the mass of the sun. Not only that, it is estimated that the size of the supermassive black hole at the center of each of those extraordinarily large galaxies can be as large as the total size of the entire solar system, based on the present observations and analyses in modern astronomy.

All in all, the take home message about the gravitational redshift caused by mechanism-revealed black holes is: a mechanism-revealed black hole can give rise to **very large** gravitational redshift in the **very big** extending region that radially spreads out from the black hole. (Related knowledge: it is estimated that there are *millions* of black holes in the Milky Way galaxy, one can thus rationally infer that a much larger number of black holes exist in many other galaxies, and that a much, much larger number of black holes, actually or practically an incalculable number of black holes, exist in the almost or virtually incalculable galaxies of the vast universe.) As a result, these numerous and massive black holes can be the main sources or dominant contributors of the gravitational redshift in the universe.

(*Friendly reminder: what this section has introduced could substantially and explicitly help one chew over, digest and comprehend the great gravitational redshift caused by the numerous and massive black holes in the universe mentioned in chapter one.)

The gravitational light bending caused by mechanism-revealed black holes

Based on the principle of gravitational light bending introduced in the last section of chapter four, all external light rays (such as from distant stars) entering the area near a (mechanism-revealed) black hole—including the region within the radius of the black hole and the hugely vast extending region outside the radius of the black hole, except a tiny fraction of (those) light rays striking on the mass in the black hole, are deflected away from the black hole by the gravitational scale contour lines of space and time in the gravitational field of the black hole (Fig. 5.3, next page). That is, all the regions affected by a mechanism-revealed black hole, except a very tiny fraction of the region occupied by the mass of the black hole, do not accept the external light rays coming towards them. And so, the take home message about the nature of the gravitational light bending caused by mechanism-revealed black holes is: **the light shines into the darkness** (the darkness is mechanism-revealed black holes), **but the darkness (virtually) does not accept it**.

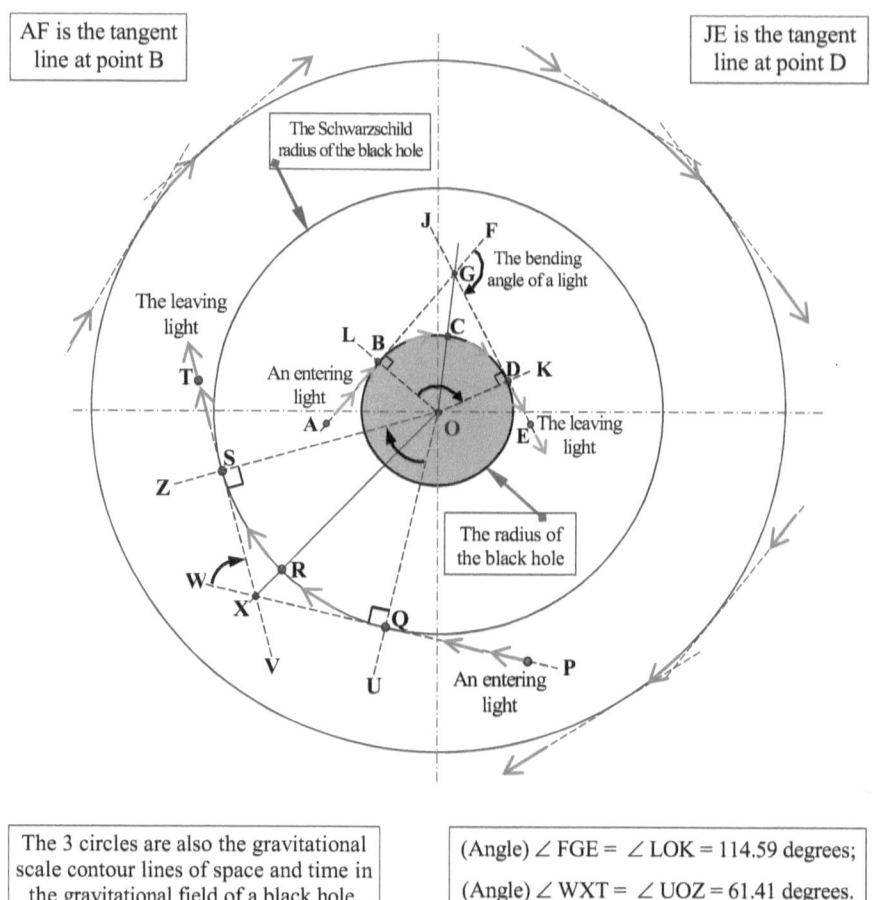

Figure 5.3, an illustration of the gravitational light bending caused by a (mechanism-revealed) black hole.

After introducing the nature of the gravitational light bending caused by mechanism-revealed black holes, let us see this nature via the following three gravitational bending angles of light with characteristic features, being calculated with the gravitational light bending equation in the newly discovered and verified gravitational theory (which is mechanism-revealed gravitational theory). (i) The gravitational bending angle of light caused by the gravitational scale contour line of space and time (which is in the gravitational field of a (mechanism-revealed) black hole) along the famous Schwarzschild radius of the black hole is 61.41 degrees (for example, the traveling route of a light along the path of P → Q → R → S → T, Fig. 5.3, P. 84; note: a gravitational scale contour line of space and time is a circle in 2-D, a sphere in 3-D). (How large is this bending angle? Please see the following two examples. A right angle is equal to 90 degrees. Each angle of an equilateral triangle, a triangle in which all three sides are equal, is equal to 60 degrees.) (ii) The gravitational bending angle of light caused by the gravitational scale contour line of space and time (which is in the gravitational field of a (mechanism-revealed) black hole) along the radius of the black hole is 114.59 degrees (for example, the traveling route of a light along the path of A → B → C → D → E, Fig. 5.3). (Related knowledge: the value of the radius of a mechanism-revealed black hole is equal to one-fourth of that of the famous Schwarzschild radius, as mentioned earlier.) (iii) The gravitational bending angle of light (caused by a (mechanism-revealed) black hole) has an upper limit or maximum value, which is equal to 229.183 degrees (that is, the gravitational bending angle of light can approach this limit, but cannot go beyond this limit). The above specific values, due to their characteristic features, can make or help one have a concrete perception or vivid impression about the nature of the gravitational light bending caused by mechanism-revealed black holes.

Last but not least, what should be pointed out is that the above three gravitational bending angles of light (61.41 degrees, 114.59 degrees, and 229.183 degrees), along with Fig. 5.3, also explicitly show that a (mechanism-revealed) black hole can cause very large gravitational light bending in the great extending region that radially spreads out from the black hole. That is, (mechanism-revealed) black holes, due to their hugely massive feature, can give rise to very large gravitational light bending in the very big extending regions that radially spread out from them.

The appendix of chapter 5

The event horizon of a black hole turns out to be an unreliable concept

As pointed out earlier in this chapter, black holes, when they are interpreted with *postulate-based* black hole theories, which are the black hole theories based on *postulate-based* general relativity, are *postulate-based* black holes (related reminder: the black holes mentioned and discussed in this appendix are *postulate-based* black holes).

In these *postulate-based* black hole theories, including the black hole theories developed by Stephen Hawking, the event horizon (the boundary of a *postulate-based* black hole) is a crucially important concept. For instance, in his book *A Brief History of Time*, Hawking pointed out that: the event horizon of a black hole is formed by the light rays that just fail to escape from the black hole. Moreover, whenever black holes are talked about, light is always an indispensable, also crucially important, subject. This prominent feature can be clearly seen from the present definitions of black holes, like: a black hole is an object so massive that even *light* cannot escape from it; if enough mass is concentrated in a small enough region, the pull of gravity in such case becomes so strong that not even *light* can escape this region; and so forth. As a result, such a prominent feature further shows or confirms that the event horizon is indeed a crucially important concept in *postulate-based* black holes thus in *postulate-based* black hole theories.

Given that all the *postulate-based* black hole theories have not solved the four fundamentally and crucially important problems in science for a very long time—these four problems are the mystery of dark matter, the mysterious sources of gamma ray bursts, the mysterious sources of ultrahigh-energy cosmic rays, and the GZK paradox (this paradox has been concisely explained in the Glossary of this book); given that the event horizon is a crucially important concept in these *postulate-based* black hole theories, it seems neither irrational nor inappropriate to inspect whether the event horizon is a reliable concept or not. (Commentator: yes, that's a good idea. In fact, it is not only rather rational but also quite wise to do such an inspection.)

Before carrying out this inspection, we need to know about the origin of the event horizon. The event horizon originates from the *momentum* of photons (momentum is a quantity of motion of a moving object), being shown in the following known procedure leading to the birth of the event horizon. First, starting with the argument like: though a photon has no rest mass, it nevertheless behaves in collision as though it has an inertial mass, *because a photon has momentum*, the same feature as the inertial mass of an ordinary object, such as a car or a bicycle (the inertial mass of an ordinary object is defined based on Newton's second law. According to this law, an object with a certain amount of mass will accelerate, or change its speed, at a rate that is proportional to the magnitude or value of the net force acting on the object; this 'a certain amount of mass' is defined as the inertial mass of the object). The assumption of 'photons have inertial mass' (in general relativity) was, therefore, proposed by ascribing the momentum of photons to their assumed inertial mass. Then, by another assumption (which is 'the equivalence of inertial and gravitational mass', also in general relativity), it was inferred that photons have gravitational mass thus being acted by gravity. Finally, the event horizon was derived from the escape velocity of an ordinary object that has a certain amount of inertial mass. (It should be mentioned that, as the mathematical reflection of the procedure above, the event horizon of a black hole was usually derived and/or explained through the following three representative steps. First, determining the escape velocity of an ordinary object having a certain amount of inertial mass, with equation A that is from Newtonian gravitation theory. Then, directly by letting this escape velocity equal the speed of light in equation A, obtain equation B. Finally, by finding the solution of equation B, obtain the event horizon, which is expressed as the radius circling around a massive celestial body, or a black hole.)

Through the pieces of information presented above, one can clearly see that the event horizon was obtained along the route: the momentum of photons → photons have inertial mass → photons have gravitational mass → photons are acted by gravity, then by replacing the escape velocity of an ordinary object that has a certain amount of inertial mass with the speed of light → the event horizon. This route shows that the *momentum* of photons is indeed the origin of the event horizon. Therefore, the task of inspecting whether the event horizon (the boundary of a black hole) is a reliable concept can be carried out through inspecting whether the current method of calculating the momentum of photons is right or wrong. That is, if this method is correct, one can say that the event horizon is a reliable concept; on the contrary, if this method turns out to be wrong, one cannot come to the conclusion that the concept of the event horizon is reliable. (Related question and answer: more specifically thus more directly, what is the relation between the concept of the event horizon and this method? Answer: the concept of the event horizon originates from the *momentum* of photons; this method determines the value of the *momentum* of photons.)

As a preparation to carry out this task, we need to know or review some basic knowledge about the momentum *change* when two objects collide (momentum is a quantity of motion of a moving object, measured as its mass multiplied by its velocity). Let us use the following situation as a simple and clear example. There are two objects, objects A and B, and immediately before their collision, *the momentum of object A is far greater than that of object B*—that is, compared to the momentum of object A, the momentum of object B is so small that it can be practically negligible. (You can think this situation as that, object A is like a racing vehicle in a sport competition running at a velocity about two hundred miles per hour, whereas object B is like a bicycle going at a velocity about ten miles per hour.) Let us check, with two different methods, the momentum of objects A and B immediately *after* their collision, in the following two cases.

First case: immediately after object A collides object B, method X calculates that the momentum of object B is more than *200 times* the original momentum of object A; method Z calculates that the momentum of object B is about 97 percent of the original momentum of object A. The related question is: even merely as a rough estimate, which method is wrong and which method is right? The answer is crystal clear: method X is definitely wrong because it incurs an obviously ridiculous result; a bit like that a 2-kilogram hen produced a 50-kilogram egg! Method Z is probably correct because its result seems to be reasonable.

Second case: immediately after object A collides object B, and under the condition that the momentum of object A has not significantly reduced in the collision (such as the momentum of object A after the collision is still about 95 percent of its original momentum), method X calculates that the momentum of object B is about *two times* the original momentum of object A; method Z calculates that the momentum of object B is about 30 percent of the original momentum of object A. Again, the related question is: even just as a rough estimate, which method is wrong and which method is right? The answer is very clear: method X is undoubtedly wrong because it incurs a clearly unreasonable result; method Z is probably correct because its result seems reasonable.

Now, let us return to our task: inspecting whether the current method of calculating the momentum of photons is correct or not. (This method was proposed by Einstein in 1916. According to this method, the momentum of a photon is equal to the energy of the photon divided by the speed of light; and the energy of the photon equals Planck constant, which is a fundamental constant in physics, multiplied by the frequency of the photon.)

One inspection is performed by an example about the photoelectric effect, being the phenomenon of ejecting electrons from a metal when it is shone by high-energy light. (The photoelectric effect tells us: when high-energy light shines on a metal surface, electrons are ejected from the metal surface. The phenomenon of the photoelectric effect was of great importance in the early 20th century for its experimentally demonstrating the particle-like nature of light, the feature simply called photons today, which are the tiny and discrete quanta or particles of light. It should be well aware that the photoelectric effect, viewed from the angle of photons, shows us such a clear fact: when a high-energy photon strikes an electron, the electron is ejected from the metal surface; this fact is explicitly reflected in the explanation of the photoelectric effect.) This example is: ultraviolet light of wavelength 150 nanometers falls on a chromium electrode (note: one thousand million (1 with nine zeros after it) nanometers = one meter, or one million nanometers = one millimeter; millimeter is usually the smallest graduation on a ruler or a meter stick).

In this example, I have calculated that, with the current method of calculating the momentum of photons, the momentum of an ejected electron is more than *240 times* the original momentum of the photon that strikes the electron. Thus being quite similar to the first case above, such a result clearly shows that this method turns out to be definitely wrong, because it incurs an obviously ridiculous result (by clearly violating momentum conservation law, one of the most fundamental principles in physics. About this law, please refer to the Glossary of this book if necessary); and again, a bit like that a 2-kilogram hen produced a 50-kilogram egg! (Related question and answer: why is the current method of calculating the momentum of photons wrong? Answer: this method greatly underestimates the momentum of photons, especially when the energy of a photon is very high—that is, the wavelength of a photon is very short, or its frequency is very high. This answer will be seen even more clearly in the next paragraph, in comparison to the result calculated with another method.)

By contrast, in the *same* example, I have also calculated that the momentum of an ejected electron is 97 percent of the original momentum of the photon that strikes the electron, using electron-photon momentum relationship (which was discovered and verified not very long ago by me). This relationship, being the combination of momentum conservation law (one of the most fundamental principles in physics) and the newly discovered and verified law of an orbiting electron with periodic impulses emitting photons (about this new physical law, please refer to the Glossary of this book if necessary), reveals the connection in momentum between an electron and its emitting photons *for the first time in the history of science*. This relationship shows that: the momentum of a photon comes from and is equal to the momentum *change* of the electron emitting the photon over this electron's one complete impulse period (in both value and direction); and the momentum of a photon is equal to the orbiting momentum of the electron emitting the photon (in value). What should be noticed here is: though electron-photon momentum relationship gives the reasonable result, the purpose of mentioning this result is to display, through a sharp comparison, that the current method of calculating the momentum of photons turns out to be clearly wrong. And this purpose should be the focus of dear readers here.

Another inspection is carried out via an example, in which a high-energy photon, such as an X-ray photon, strikes an electron that is almost stationary, which means that the momentum of the electron is much, much smaller than that of the X-ray photon striking it (that is, compared to the momentum of the X-ray photon, the momentum of the electron is small enough to be practically or safely negligible). (Such an example has realistic and factual grounds. For instance, in the experiment of the famous Compton scattering, conducted in 1922, X-rays were used to strike electrons. The Compton scattering was widely recognized as important, independent evidence supportive of the particle-like behavior of light.) In the example, I have calculated that, using the current method of calculating the momentum of photons, the momentum of an electron (after being struck by an X-ray photon) is about *two times* the original momentum of the X-ray photon that strikes the electron, whereas the X-ray photon loses only about 5 percent of its momentum in the striking (that is, the X-ray photon still keeps about 95 percent of its original momentum after the striking). So, being very similar to the second case above, this result clearly tells us that the current method of calculating the momentum of photons cannot be correct, because it leads to an obviously unreasonable result. (For comparison, in the *same* example, I have also calculated that, with electron-photon momentum relationship just mentioned above, the momentum of a struck electron is about 30 percent of the original momentum of the X-ray photon that strikes the electron.)

The explicit conclusions from these two inspections have thus uncovered or shown the irrefutable *fact* that the current method of calculating the momentum of photons turns out to be clearly wrong, because it has led to obviously ridiculous, plainly unreasonable results. This irrefutable *fact* has no choice but to tell us such a clear, unavoidable *reality:* the event horizon (the boundary of a *postulate-based* black hole) turns out to be an unreliable concept, because it originates from the momentum of photons, as analyzed and shown above; because it is this very method that determines the value of the momentum of photons, as pointed out above.

*Related and deep discussions (which can be optional, though some readers may perceive or realize that this kind of discussion turns out to be not only very interesting and inspiring, but also quite meaningful and helpful). Having known the irrefutable *fact* that the current method of calculating the momentum of photons turns out to be clearly wrong, it seems neither unusual nor unreasonable that some readers may have the possible or potential reactions like: surprised, even shocked, hard to believe, or regretted. Each of these reactions is understandable or quite normal, considering this *fact* is probably too new or too radical in their eyes. Nevertheless, is there something more important than these reactions themselves? Or more specifically, what bitter lesson can we draw from this irrefutable *fact* or from such a grave mistake? To this question, different people may have different answers. Though I firmly believe that some, even many, readers are well able to have better answers than me, I would still like to share mine with dear readers here (certainly, all readers are welcome to have comments on my answer).

My answer is: in science, it is crucially important to examine an idea or method from the different situations or processes that are covered by the idea or method, provided that these situations or processes are inherently related with each other, in order to ensure that this idea or method can give an overall coherent picture. In other words, it is fundamentally necessary to advocate and encourage the mode of comprehensively and systematically thinking in science, in order to avoid ridiculous or obvious mistakes.

This answer can be specifically reflected in the two historical events or instances. First instance: Einstein proposed the current method of calculating the momentum of photons in 1916, and he also interpreted the photoelectric effect in 1905. So if he had gone back and checked this method with the related result in the photoelectric effect, the *fact* that this method turns out to be ridiculously wrong could have been identified in the 1910s. Second instance: the famous experiment of the Compton scattering was conducted in 1922, so, and again, if Arthur Compton had checked this method with the related result in his experiment, the *fact* that this method turns out to be clearly wrong could have been detected in the early 1920s.

Yes, while there is a viewpoint or opinion saying that history has no assumptions or 'IFs', many people might have already realized or noticed that history, including the history of science of course, does have some amazing similarities. And some of these similarities have brought us some profound regrets. For example, another famous regret is that Aristotle's idea, which said that a heavy body should fall faster than a light one, had dominated the world for nearly two thousand years, no one until Galileo Galilei showed this idea turned out to be wrong. In a sense, we should not blame former generations for these regrets, instead what seems to be really or more important is that we should draw bitter lessons from these regrets. Thus, in this sense, we should not regret too much about the grave mistake in the current method of calculating the momentum of photons—this method only has a history of one century.

After the discussions above, let us still return to the concept of the event horizon. Let us go to another feature of this concept, its self-contradictory feature, because such a feature can considerably help us further think over the *reality* that the event horizon (the boundary of a *postulate-based* black hole) turns out to be an unreliable concept.

(Commentator: of course, through this self-contradictory feature, one can see this *reality* more clearly and explicitly, know this *reality* more comprehensively and thoroughly, thus comprehend this *reality* more impressively and confidently.) Then what is the behavior or nature of this self-contradictory feature?

As mentioned above, in deriving the event horizon, the escape velocity of an ordinary object with a certain amount of mass was forcibly replaced with the speed of light by making the former equal the latter. However, special relativity says that the speed of light is an unattainable speed for *any* ordinary object that has a mass, and *any* objects with mass must move at the speed less than that of light, no matter how to increase their speed. In other words, special relativity does not allow the velocity of an ordinary object having a mass to jump to the speed of light; that is, special relativity does not permit the appearance of the concept of the event horizon at all! On the other hand, this concept was developed crucially based on general relativity, which is clearly reflected in the fact that this concept is indispensably dependent on the two assumptions of general relativity (one is 'photons have inertial mass'; another is 'the equivalence of inertial and gravitational mass'), as mentioned above. That is to say, general relativity does allow the appearance of this concept. Therefore, it becomes crystal clear that the concept of the event horizon turns out to be plainly self-contradictory, when viewed from the angles of special relativity and general relativity. Undoubtedly, such a self-contradictory feature can substantially help one further realize the *reality* that the event horizon (the boundary of a *postulate-based* black hole) turns out to be really an unreliable concept.

Chapter 6
Unveil the Mystery of Dark Matter

[The Window on This Chapter]

In science, it is obviously of fundamental necessity and crucial importance to solve the problem of **dark matter** or unveil its mystery, because the amount of **dark matter** is about **5.5 times** of that of the ordinary matter in the universe (that is, about **85%** of the total matter in the universe is **dark matter**).

However, the problem of **dark matter** (or its mystery) has not only become a long-term unsolved, fundamentally important problem in science, but has also been widely recognized by the scientific community as one of the two greatest challenges to the science of the 21st century.

Only mechanism-revealed black holes (the black holes revealed and explained in chapter five) can unveil the long-standing mystery of **dark matter** (this mystery includes the constituents and fundamental nature of **dark matter**).

Dark matter is the mass in mechanism-revealed black holes; or the mass in mechanism-revealed black holes is **dark matter**.

A cluster of dark matter is a hugely massive celestial body that reduces the scales of length and time in its vicinity to such an extent that all visible light becomes invisible—all the visible light entering the region of the cluster of **dark matter** becomes invisible; all the light emitted from the cluster of **dark matter** is also invisible in the region of the cluster of **dark matter**, resulting in this region looking totally dark to the observers far away from it.

The concept of dark matter

Concisely, dark matter is a general term in astronomy and cosmology (the scientific study of the universe) used to describe the matter in the universe that cannot be seen. Though the effects of dark matter have been observed and noticed since the early 20th century, the concept of dark matter was formally proposed more than 30 years ago when it became clear that *all the galaxies* behaved as if they were far more massive than they seemed to be—without dark matter, galaxies would fly apart. Moreover, it is estimated that the amount of dark matter is about 5.5 times of that of the ordinary matter (such as the matter of stars and planets) in the universe; that is to say, about 85% of the total matter in the universe is dark matter.

While the fundamental nature of dark matter has been a thorny problem in science for more than several decades, it has been known that the existence of dark matter causes two noticeably observed gravitational effects. One is that the stars in the outer regions of a galaxy orbit *much* faster than they would if there were only ordinary matter present. (Thus, the inevitable or inescapable implication of this effect is: due to the existence of dark matter, the gravitational field over the *entire* region of a galaxy is *much* stronger than it would be if only ordinary matter existed.) Another is that the light rays from distant stars are bent (in the gravitational field of dark matter) by the enormous gravity of dark matter (such a phenomenon due to the huge gravitational effects of dark matter is referred to as the gravitational lens, being one of the main, effective and simple ways to identify the existence of dark matter nowadays). Therefore, the existence of dark matter increases gravitational effects—considerably and markedly.

Since the existence of dark matter increases gravitational effects greatly—not only considerably but also markedly; since gravitational fields are practically omnipresent in the universe (note: the universe, apart from a tiny fraction of space occupied by various celestial or heavenly bodies themselves, is composed of the numerous gravitational fields around the numerous celestial bodies such as stars, planets and black holes in each galaxy, as well as the almost or virtually incalculable gravitational fields around the almost or virtually incalculable galaxies), clearly, even obviously, in science it is fundamentally necessary and crucially important to unveil the mystery of dark matter or solve the problem of dark matter. As a matter of fact, the mystery of dark matter (or the problem of dark matter) has been fully recognized or widely accepted by the scientific community as one of the two most important problems in science (the other one is the mystery of dark energy).

Dark matter—one of the greatest puzzles in science

While the mystery of dark matter (or the problem of dark matter) has been fully recognized or widely accepted by the scientific community as one of the two most important problems in science, the harsh, also inescapable and widely admitted, *reality* is that the mystery of dark matter has been an extremely tricky problem for several decades. In fact, the long-standing mystery of dark matter has been widely acknowledged as 'one of the greatest challenges to the science of the 21st century' by the scientific community.

This harsh *reality* has been explicitly recognized in several highly authoritative sources. For instance, the fundamental nature of dark matter was listed (under "what is the universe made of?") as the number one problem among the top 25 unsolved fundamental big problems in science by *Science* Magazine (the issue of July 8, 2005, to be exact; this magazine is widely known to be one of the most authoritative, most influential and most famous publications in science). And the problem of "The Mystery of Dark Matter" was collectively listed as one of the World's 20 Greatest Unsolved Problems by more than 60 brilliant scientists—among them, 40 Nobel laureates (in the book *The World's 20 Greatest Unsolved Problems* authored by John R. Vacca). Moreover, the long-standing mystery of dark matter, or the long-standing problem of dark matter, has been listed as one of the 24 outstanding unsolved problems in the encyclopedia of physics.

Then what is the **crux** of the mystery of dark matter? This **crux** is: what is dark matter made of? Or what **constituents** is dark matter made of? (Narrator: so this **crux** clearly and explicitly tells us: the mystery of dark matter lies with the mystery in the **constituents** of dark matter; or the mystery in the **constituents** of dark matter represents the mystery of dark matter; thus, to unveil the mystery of dark matter is to reveal the

constituents of dark matter. Therefore, once we know the **constituents** of dark matter, the long-standing mystery of dark matter will be unveiled immediately; this mystery will no longer be a mystery at all. That is to say, once the **constituents** of dark matter are known, the long-standing problem of dark matter will be solved immediately.)

Then what tool is employed to work on dark matter at present? Or what theory has been hired to solve the problem of dark matter? Answer: Einstein's theory of general relativity is the main and dominant theory. That is to say, it is within the paradigm or stereotype of general relativity that the problem of dark matter has become a long-term unsolved, fundamentally important problem in science, or in astronomy and cosmology. Therefore, the so-called 'one of the greatest challenges to the science of the 21st century' is actually one of the greatest challenges to Einstein's theory of general relativity that was developed in the early 20th century; and one of these greatest challenges can include the great challenge to the *fact* that general relativity is unable to solve the most fundamental problem in front of itself (friendly reminder: as analyzed and unveiled in chapter three, general relativity is unable to solve the most fundamental problem in front of itself, which is *why* space and time are variable thus relative in a gravitational field, or *why* time runs slower and *why* length becomes shorter in such a situation). (A rational, objective commentator: since general relativity is unable to solve the most fundamental problem in front of itself; given that the problem of dark matter has not been solved within the paradigm of general relativity for quite a long time, despite the diligent and unceasing efforts of many intelligent and capable experts over the last several decades, perhaps general relativity does not have the ability to solve the problem of dark matter in truth. This is because: IF general relativity had such an ability, the problem of dark matter would have already been solved much earlier by the diligent and unceasing efforts of so many brilliant experts!)

In such a situation, it seems both rational and wise, at least neither irrational nor unwise, that we ought to or shall try the newly discovered and verified gravitational theory (which is mechanism-revealed gravitational theory, or MRGT, for short), because MRGT has solved the fundamentally important problem of *why* and how space and time are variable thus relative in a gravitational field (by revealing the mechanism of *why* and how the scales of space and time are variable thus relative due to the existence of the gravitational field)—that is, MRGT has shown us *why* time runs slower and *why* length becomes shorter in a gravitational field by revealing the mechanism behind these two *whys*, as introduced in chapter four. Most likely, MRGT, because it has solved this fundamentally important problem, also has the ability to solve the long-standing, extremely tricky problem of dark matter by revealing the constituents of dark matter. (An independent commentator: such a suggestion is obviously a rational and wise choice. In fact, it's not only rather rational but also quite wise trying a new or different method after the previous method has not solved the problem of dark matter over the last several decades. So let us give an opportunity to the new gravitational theory that has solved the fundamentally important problem of *why* and how space and time are variable thus relative in a gravitational field; let us see whether this new theory can reveal the constituents of dark matter, thus solve the problem of dark matter. **Let us be persistent towards the goal of solving the problem of dark matter; let us be flexible in selecting the methods or routes to reach this goal!**)

(The response from dark matter: I am also anxiously waiting for the new gravitational theory that has solved the fundamentally important problem of *why* and how space and time are variable thus relative in a gravitational field. Maybe, this new gravitational theory is my solution.) (The response from general relativity: it is definitely rational and unquestionably wise to give an opportunity to the new gravitational theory that has solved the fundamentally important problem of *why* and how space and time are variable thus relative in a gravitational field. Moreover, it is neither rational nor wise not to give an opportunity to this new gravitational theory.)

Then can MRGT (which has been concisely introduced in chapter four) solve the problem of dark matter by revealing the constituents of dark matter? Please allow me to unveil the answer to this question in advance very briefly here (the specific process or route leading to this answer will be introduced in this chapter after a short while): yes, it can; in fact, this new gravitational theory turns out to be the key and prerequisite to solving the problem of dark matter by revealing the constituents of dark matter. This is because only mechanism-revealed black holes can unveil the mystery of dark matter (this mystery includes the constituents and fundamental nature of dark matter) by revealing the constituents of dark matter, whereas mechanism-revealed black holes are the black holes revealed and explained by mechanism-revealed black hole theory (a new black hole theory, which has been briefly introduced in chapter five) that is based on this new gravitational theory.

Before going to the subject of unveiling the mystery of dark matter or solving the problem of dark matter, let us review the definition of a mechanism-revealed black hole introduced in chapter five, because it is mechanism-revealed black holes that have unveiled the mystery of dark matter (or solved the problem of dark matter) by revealing the constituents of dark matter, which will be introduced in this chapter after a little while. This definition is: **a mechanism-revealed black hole is a hugely massive celestial body** that reduces the scales of length and time in its vicinity to such an extent that all visible light becomes invisible—all the visible light entering the region of the black hole becomes invisible; all the light emitted from the black hole is also invisible in the region of the black hole, resulting in this region looking totally black to the observers far away from it. This definition shows and determines that the gravitational nature of mechanism-revealed black holes is fundamentally and essentially the same as other ordinary celestial bodies, even though the gravitational effects caused by mechanism-revealed black holes, due to their hugely massive feature, are much, much stronger and greater than other ordinary celestial bodies.

Because the revealing of the constituents of dark matter has been realized through revealing the mechanism and essence of dark matter, let us first go to this mechanism and essence, as well as the definition of dark matter based on this mechanism and essence.

The mechanism, essence and definition of dark matter

Now let us see the mechanism, essence and definition of dark matter. Based on the newly developed *mechanism-revealed black hole theory* (MRBHT, for short) introduced in the last chapter, when the mass of a hugely massive celestial body reduces the scales of length and time in its vicinity to such an extent that all visible light becomes invisible, a black hole is thus formed; the mass of the hugely massive celestial body thereby becomes a cluster of dark matter. Therefore, the *mechanism* of dark matter is the mechanism of MRBHT or the mechanism of mechanism-revealed black holes revealed and explained by MRBHT (the *mechanism* of MRBHT is that the mass of a hugely massive celestial body reduces the scales of length and time in its vicinity to such an extent that all visible light becomes invisible, being clearly reflected in the definition of a mechanism-revealed black hole just reviewed above; the *mechanism* of MRBHT has also been introduced in the second section of chapter five). And the *essence* of dark matter is the essence of MRBHT or the essence of mechanism-revealed black holes revealed and explained by MRBHT (which shows that the *essence* of a mechanism-revealed black hole is the tremendous reduction in the scales of length and time that occurs in the gravitational field of (and near) a hugely massive celestial body. This essence has been mentioned in the second section of chapter five).

The mechanism and essence of dark matter shows and determines that the concise definition of dark matter is: **dark matter is the mass in (mechanism-revealed) black holes; or the mass in (mechanism-revealed) black holes is dark matter**. That is, dark matter is the same thing as the mass in (mechanism-revealed) black holes.

Such a concise definition of dark matter clearly indicates that the definition of a mechanism-revealed black hole is completely applicable to a cluster of dark matter (as just mentioned above, this definition is: **a mechanism-revealed black hole is a hugely massive celestial body** that reduces the scales of length and time in its vicinity to such an extent that all visible light becomes invisible—all the visible light entering the region of the black hole becomes invisible; all the light emitted from the black hole is also invisible in the region of the black hole, resulting in this region looking totally black to the observers far away from it). As a result, the complete definition of dark matter is: **a cluster of dark matter is a hugely massive celestial body** that reduces the scales of length and time in its vicinity to such an extent that all visible light becomes invisible—all the visible light entering the region of the cluster of dark matter becomes invisible; all the light emitted from the cluster of dark matter is also invisible in the region of the cluster of dark matter, resulting in this region looking totally dark to the observers far away from it. (Related reminder: this definition of dark matter shows and determines that the gravitational nature of dark matter is fundamentally and essentially the same as other ordinary celestial bodies, even though the gravitational effects caused by a cluster of dark matter, due to its hugely massive feature, are much, much stronger and greater than an ordinary celestial body.)

Most of all, the mechanism, essence and definition of dark matter above provide the fundamental, complete and thorough explanation for the above-mentioned two noticeably observed gravitational effects caused by dark matter. One is that the stars in the outer regions of a galaxy orbit *much* faster than they would if there were only ordinary matter present (this observed gravitational effect clearly indicates that due to the existence of dark matter, the gravitational field over the *entire* region of a galaxy is *much* stronger than it would be if only ordinary matter existed), because the amount of dark matter in the universe is about **5.5 times** of that of the ordinary matter (such as the matter of stars and planets)—that is, about **85%** of the total matter in the universe is dark matter. Another is that the light rays from distant stars are bent (in the gravitational field of dark matter) by the enormous gravity of dark matter (such a phenomenon due to the huge gravitational effects of dark matter is referred to as the gravitational lens, being one of the main, effective and simple ways to identify the existence of dark matter nowadays), because such a noticeably observed gravitational effect further substantiates or confirms the explicit existence of a huge amount of dark matter in the universe.

Reveal the four fundamental natures of dark matter

After introducing the mechanism, essence and definition of dark matter, we will go to the four fundamentally and crucially important natures of dark matter revealed by this mechanism, essence and definition. These four natures are: the constituents of dark matter, dark matter can emit light—even a very large amount of light rays with great intensity and high density, the great gravitational redshift caused by dark matter, and the gravitational light bending caused by dark matter. Because these natures are of fundamental and crucial importance, let us go over them one by one in the following four subsections, specifically and concisely.

Reveal the constituents of dark matter

First of all and most of all, the mechanism, essence and definition of dark matter presented above reveal the constituents of dark matter. Please notice that, as revealed by the new black hole theory (which is mechanism-revealed black hole theory, MRBHT, for short), the decisive difference between (mechanism-revealed) black holes and other ordinary celestial bodies (such as stars and planets) lies with the hugely massive feature of (mechanism-revealed) black holes, rather than in their constituents. And so, in comparison with other ordinary celestial bodies, the constituents of mechanism-revealed black holes are not fundamentally different at all—neither mysterious nor unique in essence; that is to say, fundamentally speaking, the constituents of mechanism-revealed black holes are the same as other ordinary celestial bodies. In other words, a (mechanism-revealed) black hole can be formed as a result of many ordinary celestial bodies joining together.

For instance, according to MRBHT, as long as the mass of a celestial body is greater or equal to several dozens of times the mass of the sun, the celestial body is or becomes a (mechanism-revealed) black hole, if the average density of the black hole is set at the density of an atomic nucleus. The available knowledge about the mass of various celestial bodies shows that such a requirement on mass is not difficult to meet even in (our) the Milky Way galaxy, needless to say in the vast universe. Therefore, the fundamental elements that make up the constituents of dark matter, or the fundamental constituents of dark matter, are the same as those of other ordinary celestial bodies, because **dark matter is the mass in (mechanism-revealed) black holes**; and because **a (mechanism-revealed) black hole is a hugely massive celestial body** that …. As a result, MRBHT unveils the mystery of dark matter by revealing its constituents (reminder: as analyzed and pointed out earlier in this chapter, once we have known the constituents of dark matter, the mystery of dark matter is unveiled).

*Commentator: after unveiling the mystery of dark matter by revealing its constituents, some readers might be amazed, even astonished: how simple the constituents of dark matter are! How simple the mystery of dark matter is! The mystery of dark matter, once unveiled, turns out to be amazingly simple, unbelievably simple! Yes, it is indeed. In fact, what makes science really wonderfully beautiful and fascinatingly attractive lies with that the major principles (or the major, fundamental truths) revealed by it are amazingly simple; there are quite a lot of such instances in science.

Newton's second law is simple, but many people admire and appreciate its wonderful beauty and fascinating attraction after knowing its profound implications (this law is the famous $F = m\mathbf{a}$, where m is the mass of an object, **a** is the acceleration of the object, and F is the net force acting on the object). Einstein's famous mass-energy equivalence equation is simple, but many people admire and appreciate its wonderful beauty and fascinating attraction after realizing its profound implications (this equation is $E = mc^2$, where c is the speed of light, m is the rest mass of an object, and E is the rest energy of the object). The famous Planck-Einstein equation is simple, but many people admire and appreciate its wonderful beauty and fascinating attraction after perceiving its profound implications (this equation is $E = hf$, where h is Planck constant, f is the frequency of a photon, and E is the energy of the photon). The structure of DNA is simple, but many people admire and appreciate its wonderful beauty and fascinating attraction after recognizing its great significance ... and so on.

Not only that, one of the most noticeable marks or logos in science is: wonderfulness comes from simplicity; simplicity originates from the major, fundamental truths revealed by science—'the major, fundamental truths are simple' has become one of the well-known, also widely recognized, facts or rules. Moreover, the typical or essential nature of the major, fundamental truths is simple, which has been clearly substantiated or impartially witnessed by a lot of such instances in science, like those just mentioned. In addition, according to Laozi (also called Lao-tzu), a great and famous philosopher in the Spring and Autumn Period (a period in ancient China, from approximately 771 to 476 BC), the founder of Taoism, 'All things obey or follow their own rules, but the greatest truths are the simplest'.

With the above famous instances or facts as an inspiration or encouragement, the simplicity in the constituents of dark matter seems to be, at least in an intuitive sense, a possible indication of the great wonderfulness. Of course, what really makes the revealing of the constituents of dark matter truly wonderful and attractive lies with: this revealing is the unveiling of the great mystery of dark matter (related reminder: as mentioned earlier in this chapter, the crux of the mystery of dark matter clearly and explicitly tells us that the mystery of dark matter lies in the mystery in its **constituents**. And so, once we know the **constituents** of dark matter, the mystery of dark matter will be unveiled immediately; the mystery of dark matter will no longer be a mystery at all). Moreover, the revealing of the **constituents** of dark matter is the prerequisite and key to uncovering the other three closely related, also crucially or remarkably important, natures of dark matter (these natures will be introduced in the following three subsections).

*Related clarification or relevant reminder! On the contrary, according to *postulate-based* black holes, which are the black holes interpreted by the *postulate-based* black hole theories based on *postulate-based* general relativity, the constituents of black holes have remained a great mystery. In reality, within the paradigm or stereotype of *postulate-based* black holes, the constituents of black holes have literally become one of the long-term unsolved, fundamentally important, also extremely tricky, problems in science—in physics or in astrophysics, to be exact (astrophysics is a branch of astronomy that deals with the physical and chemical structure of the stars, planets, etc.). In fact or in truth, also in effect, within this paradigm or stereotype, it is not possible to know the constituents of black holes at all. Consequently, also unavoidably, within the paradigm or stereotype of *postulate-based* black holes, it is definitely impossible to know the constituents of dark matter; this is because, as presented in the section above, **dark matter is the mass in *mechanism-revealed* black holes** that are revealed and explained with ***mechanism-revealed* black hole theory** (which is a fundamentally new black hole theory; this new black hole theory has been introduced in chapter five).

Related questions and answers following the related clarification or relevant reminder above. Question: what is the most fundamental reason or cause that the *postulate-based* black hole theories, which are based on *postulate-based* general relativity, turn out to be the dead alley or impassable obstacle to revealing the constituents of dark matter? Answer: because *postulate-based* general relativity is unable to solve the most fundamental problem in front of itself, which is *why* space and time are variable thus relative in a gravitational field, or *why* time runs slower and *why* length becomes shorter in such a situation (this inability has been explicitly mentioned in chapter three). Question: then what is the explicit or noticeable, also undeniable or irrefutable, hard evidence that unavoidably points to this inability of *postulate-based* general relativity? Answer: such evidence is the plain *truth* that *postulate-based* general relativity **absolutely and **desperately** necessitates its <u>indispensable</u> postulate of 'invariant scales of length and time', which says that the scales of length and time at different points over an entire gravitational field are the same. One can easily and clearly understand this hard evidence via the thinking like: if *postulate-based* general relativity had been able to solve this most fundamental problem, this <u>indispensable</u> postulate would not have been necessary at all, thus would never have appeared at all; moreover, if *postulate-based* general relativity had been able to solve this most fundamental problem, who would have looked for trouble by proposing such a totally redundant, completely unnecessary postulate?!? And so, this plain *truth* is actually and exactly the hard evidence that clearly and unavoidably shows this inability of *postulate-based* general relativity. As to the detailed analysis and specific discussion about this hard evidence, please go back to the related section of chapter three if necessary.

Question: what is the direct reason or cause that *postulate-based* general relativity is unable to solve the most fundamental problem in front of itself, which is *why* space and time are variable thus relative in a gravitational field, or *why* time runs slower and *why* length becomes shorter in such a situation? Answer: because *postulate-based* general relativity is unable to reveal the <u>mechanism</u> behind its fundamentally important postulate, the postulate of 'equivalence principle'; this postulate says that gravitational force has the same effect in increasing the velocity of an object as other traditional forces. What should be pointed out is that this postulate is crucially indispensable to general relativity—without this postulate, there would have been no general relativity at all; in fact, this postulate is the heart and soul of general relativity. Question: how do you know this inability of *postulate-based* general relativity? Answer: after revealing the <u>mechanism</u> behind this postulate. Question: then what is the <u>mechanism</u> behind this postulate? Answer: the <u>mechanism</u> behind this postulate turns out to be the mass consumption caused by mass doing positive work under the action of acceleration (increase of velocity), either due to gravitational force or due to other traditional forces (note: the mass consumption has been introduced in the last subsection of the appendix of this chapter). The core point of this mechanism is: the mass of an object is consumed (becoming less) due to its doing positive work, when the object is accelerated, no matter whether the acceleration is caused by gravitational force or by other traditional forces. The key to grasping this mechanism is: having revealed the mechanism behind this postulate is the explicit and sufficient evidence that there is this mechanism, a bit like: having found the continent of North America is the irrefutable evidence that there is this continent. On the other hand, general relativity is unable to reveal the mechanism behind this postulate, being a self-evident or actually admitted *fact*, thus also being an irrefutable or undeniable *fact*. (One can clearly and easily realize this inability and this *fact* via the simple and straightforward thinking like: if the mechanism behind a postulate had been revealed, the postulate would no longer have been a postulate at all; along with the specific and constant reminder from such an unavoidable *reality:* within the paradigm or stereotype of general relativity, this postulate is **always** a postulate!)

Question: what is the root cause or in-depth reason that *postulate-based* general relativity is unable to reveal the mechanism behind the postulate of 'equivalence principle'? Answer: the law of mass doing work had not been discovered before *postulate-based* general relativity was developed, because the mass consumption is the direct product of an application of the law of mass doing work, as pointed out in the last subsection of the appendix of this chapter; because the mass consumption has been proven to be the mechanism behind this postulate. (Friendly reminder: for the specific and detailed knowledge about the law of mass doing work and the mass consumption, please see the appendix of this chapter.) Thus one should not criticize *postulate-based* general relativity for having not revealed the mechanism behind this postulate, because the law of mass doing work came to the world about one century later than *postulate-based* general relativity.

Dark matter can emit a large amount of light

The second fundamental, also crucial, nature of dark matter is that dark matter can emit light, even a very large amount of light rays with great intensity and high density (light rays are also called photons), because **dark matter is the mass in mechanism-revealed black holes,** as unveiled and pointed out in the section above; and because **a mechanism-revealed black hole**, as revealed and emphasized in chapter five, can emit light, even a very large amount of light rays with great intensity and high density, from the region within its radius, though the emitted light is invisible in the region within the radius of the black hole (this is because **a mechanism-revealed black hole is a hugely massive celestial body** that reduces the scales of length and time in its vicinity to such an extent that all visible light becomes invisible—all the visible light entering the region of the black hole becomes invisible; all the light emitted from the black hole is also invisible in the region of the black hole, resulting in this region looking totally black to the observers far away from it). As a result, this fundamental, also crucial, nature of dark matter shows that mechanism-revealed black hole theory is the key to solving the four long-standing, fundamentally important problems in science: dark matter, gamma ray bursts, ultrahigh-energy cosmic rays, and the GZK paradox (this paradox has been concisely explained in the Glossary of this book).

(*!*In contrast, according to the *postulate-based* black hole theories based on *postulate-based* general relativity, a *postulate-based* black hole cannot emit light from the region within its boundary, the so-called event horizon, because the definition of the event horizon is: the event horizon of a black hole is formed by the light rays that just fail to escape from the black hole. However, the event horizon of a black hole (the boundary of a black hole) turns out to be an unreliable concept. Why? Please refer to the appendix of chapter five (P. 86 ~ 94).

Consequently, also unavoidably, these *postulate-based* black hole theories are not only unable to solve the four long-standing, fundamentally important problems just mentioned above, but have also become the shackles and obstacles to solving them.)

The great gravitational redshift caused by dark matter

The third fundamental nature of dark matter is the great gravitational redshift caused by dark matter. This fundamental nature of dark matter is the same thing as the nature of the great gravitational redshift caused by mechanism-revealed black holes, because **dark matter is the mass in mechanism-revealed black holes** (or the mass in mechanism-revealed black holes is dark matter), as unveiled and pointed out in the section above. And so, the explanation and description about the nature of the gravitational redshift caused by mechanism-revealed black holes, having been presented in chapter five, are completely applicable to the nature of the gravitational redshift caused by dark matter, after replacing mechanism-revealed black holes with dark matter, or replacing a mechanism-revealed black hole with a cluster of dark matter. (*Friendly reminder: this fundamental nature of dark matter can substantially and explicitly help one chew over, digest and comprehend the great gravitational redshift caused by dark matter mentioned in chapter one.)

The gravitational light bending caused by dark matter

The fourth fundamental nature of dark matter is the gravitational light bending caused by dark matter. This fundamental nature of dark matter is the same thing as the nature of the gravitational light bending caused by mechanism-revealed black holes, because **dark matter is the mass in mechanism-revealed black holes** (or the mass in mechanism-revealed black holes is dark matter), as unveiled and pointed out in the section above. And so, the explanation and description about the nature of the gravitational light bending caused by mechanism-revealed black holes, having been presented in the last section of chapter five, are completely applicable to the nature of the gravitational light bending caused by dark matter, after replacing mechanism-revealed black holes with dark matter, or replacing a mechanism-revealed black hole with a cluster of dark matter.

What should be mentioned is that, as unveiled and pointed out in in the last section of chapter five, (mechanism-revealed) black holes, due to their hugely massive feature, can give rise to **very large** gravitational light bending in the **great** extending regions that radially spread out from them. That is to say, a cluster of dark matter can cause **very large** gravitational light bending in the **great** extending region that radially spreads out from the cluster of dark matter, because dark matter is the mass in mechanism-revealed black holes (or the mass in mechanism-revealed black holes is dark matter). This gives the specific and explicit, also fundamental and thorough, explanation to one of the above-mentioned two noticeably observed gravitational effects caused by dark matter, which is that the light rays from distant stars are bent (in the gravitational field of dark matter) by the enormous gravity of dark matter (such a phenomenon due to the huge gravitational effects of dark matter is referred to as the gravitational lens, being one of the main, effective and simple ways to identify the existence of dark matter nowadays).

*Commentator: after the four fundamental and crucial natures of dark matter have been revealed, dark matter is unfolding its profound, magnificent, and wonderful beauty and charm by displaying its splendid and superb features to us! (Narrator: if the virtually innumerable clusters of dark matter in the universe, *before* these fundamental and crucial natures are revealed, are analogous to the numerous tender buds in an exceptionally large garden about to open and bloom, then these virtually innumerable clusters of dark matter, *after* these fundamental and crucial natures are revealed, become the numerous blossoming, colorful, beautiful and wonderful flowers. These beautiful and wonderful flowers of the incalculable clusters of dark matter are blossoming to you, to me, and to all of us; these beautiful and wonderful flowers are unfolding and displaying the splendid, marvelous, and astonishing beauty and charm of dark matter to our entire human beings.)

The Bitter lessons after solving the problem of dark matter

After solving the problem of dark matter (or after unveiling the mystery of dark matter) by revealing the constituents and fundamental nature of dark matter, it turns out that mechanism-revealed black hole theory (MRBHT, for short), through *mechanism-revealed* black holes—the black holes revealed and explained by MRBHT, is the key (and the only key) to solving the problem of dark matter (or unveiling the mystery of dark matter). And so, it turns out that the key (also the crucial precondition) to solving the problem of dark matter lies in finding out the new gravitational theory that is able to solve the fundamentally important problem of *why* and how space and time are variable thus relative in a gravitational field, i.e., that is able to show us *why* time runs slower and *why* length becomes shorter in a gravitational field by revealing the mechanism behind these two *whys*. This is because MRBHT is based on the newly discovered and verified gravitational theory (which is mechanism-revealed gravitational theory, having been briefly introduced in chapter four) that has solved this fundamentally important problem by revealing *why* and how the scales of space and time are reduced in a gravitational field, i.e., that has shown us *why* time runs slower and *why* length becomes shorter in a gravitational field (by revealing the mechanism behind these two *whys*). Therefore, the prerequisite (or the absolutely necessary requirement) for solving the problem of dark matter lies with having this fundamentally important problem solved.

Such a prerequisite explicitly shows and unavoidably points to: **any** *postulate-based* black hole theories, as long as they are based on *postulate-based* general relativity, never have the ability to solve the problem of dark matter or unveil the long-standing mystery of dark matter at all. This is because *postulate-based* general relativity, as analyzed and unveiled in chapter three, is unable to solve the most fundamental problem in front of itself, which is *why* space and time are variable thus relative in a gravitational field, or *why* time runs slower and *why* length becomes shorter in such a situation; due to this inability, general relativity doesn't have the ability to solve the fundamentally important problem of *why* and how space and time are variable thus relative in a gravitational field, of course. Consequently, also unavoidably or inevitably, within the paradigm or stereotype of these *postulate-based* black hole theories, any efforts attempting to solve the problem of dark matter have been fruitless and will continue to be fruitless, have been unsuccessful and will continue to be unsuccessful. (A rational, objective and independent commentator: of course, some respected, related experts might/could have the power or influence to continue working within the paradigm of these *postulate-based* black hole theories, if they are generous and brave enough to continue to waste their precious time and efforts on this definitely fruitless and unsuccessful paradigm. However, I do hope that these experts would/could truly treasure or really cherish their precious time and efforts.)

(*Commentator: the information presented in the two paragraphs above is clearly rational and noticeably reasonable, thus quite easy to understand, because this information simply tells us that solving the problem of dark matter, or unveiling the mystery of dark matter, is ultimately dependent on the theory that is able to tell us *why* time runs slower in a gravitational field, rather than on the theory that is unable to.)

This very prerequisite also teaches us such a bitter lesson or a painful experience: in science there are no shortcuts at all; any theory shall not detour or skip over the most fundamental problem in front of itself; any theory should solve the most fundamental problem it has to face! If a theory detours or skips over the most fundamental problem in front of itself, the theory may subsequently bring about some extremely serious consequences or troubles; the theory may even give rise to endless troubles for the future.

The appendix of chapter 6

The Law of Mass Doing Work—about Why Mass Has Energy

<The key functions or core roles of the law of mass doing work>

The law of mass doing work, because it shows *why* mass has energy by revealing that mass has the ability of doing work, is one of the most fundamental and most important laws in science.

The law of mass doing work, being a newly discovered and verified physical law, reveals and shows: when a mass does positive work, the mass thus decreases; but when a mass does negative work, the mass thus increases.

The law of mass doing work is the *only* physical law that has revealed the mechanism of/behind the famous mass-energy equivalence equation; and *only* this mechanism can answer the biggest *why* underlying this great equation, *why* mass has energy— because mass has the ability of doing work.

The law of mass doing work is indispensable for unveiling the long-standing mystery of dark matter (this mystery includes the constituents and fundamental nature of dark matter), because *only* this law can lay the fundamental theoretical basis of the new theory that is onto the track to unveil this mystery.

The questions pointing to the law of mass doing work

Question one. The famous mass-energy equivalence equation (which is $E = mc^2$, where c is the speed of light, m is the rest mass of an object, and E is the rest energy of the object) tells us that mass and energy are equivalent. Since energy can do work—being a universally accepted *fact*, why can't mass? More specifically thus to be more perceptible or noticeable, since energy has the ability of doing work—being a fully recognized *fact*, and since mass and energy are equivalent, it is clear, even obvious, that mass also has the ability of doing work; it is necessary and inevitable that mass has the ability of doing work. Therefore, the famous mass-energy equivalence equation clearly and definitely points to the law of mass doing work (because only this law can reveal, express and reflect the ability of mass doing work).

Question two. To the fundamentally important question in science or in physics: *why* does mass have energy? The answer from the law of mass doing work is: because mass has the ability of doing work. In fact, only this law can answer such a big question, because only this law can reveal, express and reflect the ability of mass doing work. Therefore, this fundamentally important question explicitly and unmistakably points to the law of mass doing work.

Question three. One can think over the law of mass doing work conversely if necessary: if mass could not have the ability to do work, then it would be scientifically groundless to say that mass has energy! Or what would be the scientific basis to say that mass has energy if mass could not have the ability to do work? (Commentator: yes, that's correct. If mass could not have the ability to do work, then the well-known *fact* that mass has energy would actually become totally groundless or baseless in truth. If mass could not have the ability to do work, the fully acknowledged, also widely accepted, concept of 'mass-energy equivalence' would inevitably lose its most fundamental, most essential and most important implication; this concept would thus become meaningless.) And so, the well-known *fact* that mass has energy has no choice but to tell us another clear *fact:* mass has the ability of doing work. As a result—as an explicit and noticeable result in essence, also as an undeniable or irrefutable result in truth, this clear *fact* evidently and inevitably points to the law of mass doing work (because only this law can reveal, express and reflect the ability of mass doing work).

All in all, the above three questions collectively and consistently, also explicitly and undeniably, point to the objective and real existence of the law of mass doing work. And so, if this law has been discovered, such a discovery ought to be readily accepted.

The law of mass doing work

The law of mass doing work came to the world recently, because it was discovered and verified not very long ago by me. The core principle of the law of mass doing work is: the amount of energy in the mass of an object is measured and determined by the amount of work done by the object's mass. (Narrator: such a core principle, when viewed from the angle of comprehension, is quite comparable or very similar to that the energy of a body is measured and determined by the body's ability of doing work in classical physics. Quite obviously, such a noticeable comparability or similarity is a substantial help for one to perceive and grasp this core principle easily and quickly.)

Concisely, as the exact reflection of this core principle, the law of mass doing work (with accurate mathematical expression) shows that, when the velocity of an object is increased, the object's mass does positive work, the object thus loses the same amount of energy as that of the work done by the mass of the object from and by consuming its mass. As a result, the core point of this law is: an object's mass doing positive work causes a corresponding decrease in the object's mass; that is, when a mass does positive work, the mass thus decreases. (One can clearly and easily understand this core point via such a simple comparison in classical physics: when a body does positive work, the energy of the body decreases.) The other side of this core point is: an object's mass doing negative work, which occurs when the velocity of the object is decreased, causes a corresponding increase in the object's mass; that is, when a mass does negative work, the mass thus increases. (One can clearly and easily understand this side via such a simple comparison in classical physics: when a body does negative work, the energy of the body increases.)

*Friendly reminder: dear readers, if you are the professional people in physics, especially in modern physics, you can comprehend the law of mass doing work more easily and quickly than others. This is because the law of mass doing work directly and totally comes from the relativistic kinetic energy of an object with rest mass m (by exchanging the positions of the velocity v and the relativistic momentum p in the integral calculation of the relativistic kinetic energy of the object, then by the definite integral operation from 0 to v with velocity v as variable). As a result, the relativistic kinetic energy (of the object) is the area that is *under* the line of a certain velocity v and *above* the line of the relativistic momentum p; whereas the law of mass doing work is the area that is *under* the line of the relativistic momentum p. That is to say, the law of mass doing work is not only connected to but also based on the known or conventional knowledge; such a feature can substantially enhance the recognition and acceptance of this law, though it is unconventionally new. Moreover, the law of mass doing work can be simply expressed as the product of the force acting on an object with rest mass m and the displacement of the object; i.e., this expression is totally consistent with the fully recognized expression of the work done by a force in classical physics. Quite obviously, also rather rationally, such a consistency can not only substantially but also explicitly enhance the recognition and acceptance of this law, even though it is a newly discovered physical law.

What should be pointed out is that the concept and equation of the relativistic kinetic energy of an object with rest mass m has been written into the textbooks for college or graduate education (since far more than half a century ago); that is, this concept and equation has already been fully recognized and accepted by the scientific community in physics. And so, the professional people in physics or in modern physics, because they have been familiar with this concept and equation very well, really have an obvious advantage over others in comprehending the law of mass doing work, which can make them comprehend this law much more easily and quickly than others, even though the discovery of this law might be radically new in the eyes of some of those respected conventional professional people. So my sincere congratulations go to the professional people in physics for this obvious advantage; please accept my sincere and rational congratulations if you are the professional people. (An independent commentator: because the law of mass doing work directly and entirely comes from the equation of the relativistic kinetic energy of an object with rest mass m; because this equation has been fully recognized and completely accepted by the scientific community in physics; because nothing is added or removed in the process from this equation to the law of mass doing work, it seems that there is neither rational reason nor valid basis not to accept the law of mass doing work. In fact, there is utterly no way to deny the law of mass doing work from the angle of science.)

The mechanism of the mass-energy equivalence equation

As a direct application of the law of mass doing work, this law has revealed the mechanism of/behind the famous mass-energy equivalence equation (this equation is $E = mc^2$, where c is the speed of light, m is the rest mass of an object, and E is the rest energy of the object. This famous equation is often referred to as the greatest equation in the history of science nowadays).

This mechanism turns out to be: the rest energy of an object, being the total energy contained in the rest mass of the object, is equal to the maximum ability of the object's mass doing positive work (this maximum ability is equal to mc^2, the right side of the famous mass-energy equivalence equation). So this mechanism shows that the world-famous mc^2 (in this famous equation) turns out to be not only the total energy contained in a rest mass m, but also the maximum ability or capacity of the rest mass m doing (positive) work. Moreover, this mechanism further shows that the very reason *why* the total energy contained in a rest mass m is equal to mc^2 is because the maximum ability or capacity of the rest mass m doing (positive) work is equal to mc^2.

After revealing the mechanism of/behind this famous and great equation, the solid existence of this mechanism is an irrefutable fact, a bit like: having found out the continent of North America is the irrefutable fact that there is this continent. More than that, the validity and reliability of this great equation thus become further solid and secure, after finding out its mechanism—because it turns out that this great equation does have a very solid and secure mechanism. (Related question and answer: what is the big or essential difference *before* and *after* revealing this mechanism? Answer: before this revealing, people merely knew mass has energy, but couldn't know *why;* after this revealing, people know *why* mass has energy via knowing that mass has the ability of doing work. Accordingly, this big or essential difference is also an explicit demonstration of the fundamental importance of the law of mass doing work, corresponding to the fundamentally important status of this famous and great equation in science.)

(Commentator: wow! Aha! The mechanism of/behind this famous and great equation is finally brought to light. This is definitely a remarkably important event to this great equation, because this mechanism, only this mechanism, is able to answer the biggest *why* underlying this great equation, *why* mass has energy—because mass has the ability of doing work! Moreover, if one thinks over this mechanism for a few minutes, it seems not difficult that he or she can clearly and surely realize: only the law of mass doing work is able to reveal the mechanism of/behind this great equation, believe it or not. This realization can easily enable one to be aware: the existence of this law turns out to be an explicit *fact*, a clear *fact*, also an undeniable or irrefutable *truth*, because the existence of this famous and great equation has become a universally acknowledged, well-known *fact;*

because only this law can reveal the mechanism of/behind this very equation. Thus, even if Bingcheng Zhao, the author of this book, had not discovered this law, somebody else would find it someday, sooner or later—of course; the earlier, the better. Yet regardless of who has discovered this law, the famous mass-energy equivalence equation is or should be equally happy, because this law has revealed the mechanism of/behind this famous and great equation, which is also often referred to as the greatest equation in the history of science nowadays. After this revealing, this famous and great equation, via unfolding and displaying its mechanism, appears even more beautiful and charming.)

When the famous mass-energy equivalence equation, or the greatest equation in the history of science, is at the age of more than one hundred years old, its secret veil is finally unveiled—the mechanism of/behind this famous and great equation has been at last revealed. (Reviewer: fortunately, this famous and great equation is not a bride! Of course, if it had been a bride, probably no bridegroom in the world would have been patient enough to wait for such a long time to unveil her veil after their wedding. But for a fundamentally important and extraordinarily influential equation in science like this famous and great equation, it seems that the later unveiling its secret veil, the more wonderful its marvelous charm is. This seems to be a bit like wine: a bottle of old wine is more tasteful and mellower than a new one.) In this sense, the famous and great mass-energy equivalence equation should be happy for itself—be happy for its mechanism having been finally revealed. In this sense, this famous and great equation ought to congratulate on itself—congratulate on its mechanism having been at last revealed. In this sense, this famous and great equation must celebrate itself—celebrate its mechanism having been finally brought to light! Moreover, and in a broad sense, it seems acceptable if this famous and great equation wants to invite all the people in the world, especially the respected and related experts in physics, to have a grand

and solemn celebration of this great and historic revealing! (Most probably, this famous and great equation will provide delicious food and excellent wine for all of us in such a spectacular, splendid, and wonderful occasion.)

What should be mentioned is that the above-mentioned fact, which is that the law of mass doing work has revealed the mechanism of/behind the famous mass-energy equivalence equation, is also fundamentally and crucially important to this newly discovered law. This fundamental and crucial importance is reflected in such a clear *fact:* this law has been verified or confirmed via its revealing the mechanism of the famous mass-energy equivalence equation, because this famous equation has passed experimental tests many times since its birth, thus being a fully recognized and universally accepted *fact*. With this verification or confirmation, the validity and reliability of this law are thus quite positive. (So, for this verification or confirmation, this law, on behalf of me, wants to express its sincere acknowledgment to all the related scientists for their great contributions that have made this famous and great equation found and verified, especially to the great scientist Albert Einstein, the founder of this equation.) (Related question and answer: because the law of mass doing work has revealed the mechanism of/behind this famous and great equation, can this very equation itself be a good window, through which one could see this law more clearly, thus have a better understanding of this law? Answer: yes, it can; please see the specific and closely related facts or information in the coming paragraph.)

One could tangibly and quickly grasp the law of mass doing work if he or she views this law from the following several important, also easily perceptible, angles. Angle A, from the large perspective of the fundamental question: why does mass have energy? The answer from this law is: because mass has the ability of doing work; in fact, only this law can answer such a fundamental question. Angle B, the famous mass-energy equivalence equation tells us that mass and energy are equivalent; and since energy can do work—being a universally accepted *fact*, why can't mass? More specifically, since energy has the ability of doing work—being a fully recognized *fact*, and since mass and energy are equivalent, it is clear, even obvious, that mass also has the ability of doing work; it is necessary and inevitable that mass has the ability of doing work. (Otherwise, the fully acknowledged concept of 'mass-energy equivalence' would lose its most fundamental, most essential and most important implication; this concept would thus become meaningless in fact.) Angle C, one can think over this law conversely if necessary: if mass could not have the ability to do work, it would be scientifically groundless to say that mass has energy! In other words, the known *fact* that mass has energy has no choice but to tell us another clear *fact:* mass has the ability of doing work. Angle D, since mass has the ability of doing work, it becomes quite natural that, when an object's mass does *positive* work, the object's mass thereby *decreases* (one can clearly perceive and easily comprehend this point if he or she is familiar with such common knowledge in classical physics: when a body does *positive* work, the available energy of the body thus *decreases*). (Commentator: when one views the law of mass doing work through these diverse visual angles, he or she could see this law more clearly from different directions, a bit like 3-D visual effects; he/she could thus grasp this law more tangibly and effectively. Once one has grasped this law, he or she could clearly and easily realize that this law can lay the solid foundation for any theories based on it.)

More than what we have seen above, the very fact, which is that the law of mass doing work has revealed the mechanism of/behind the famous mass-energy equivalence equation, clearly and definitely points to the great importance of this law along the following explicit and noticeable direction. Since this famous equation is widely recognized as the greatest equation in science, then its mechanism is, or ought to be, the greatest mechanism in science; since the law of mass doing work is the only physical law that reveals this greatest mechanism, then it is actually rational and appropriate (or at least it is neither irrational nor inappropriate) if one comes to the conclusion that this law is the greatest law or one of the most fundamental and most important laws in science. (Commentator: yes; this conclusion is obviously rational and objective, thus appropriate. And so, it is no exaggeration to say that the discovery of the law of mass doing work is really and truly a great achievement or historic event in science, believe it or not.) (Correspondingly, what readers have seen above is, or ought to be, the greatest law or one of the most fundamental and most important laws in science; so the author of this book genuinely congratulates on you, dear readers.)

The mass consumption

The mass consumption (being the direct result of the combination of the two things just mentioned above: the law of mass doing work and the mechanism of the famous mass-energy equivalence equation revealed with this law) shows that the mass of an object *decreases* with the increase in its velocity, by revealing *why* and *how* the object's mass is being consumed due to its doing positive work. Therefore, the mass consumption, being caused by mass doing positive work, is simply the product of an application of the newly discovered and verified law of mass doing work.

One could clearly and easily understand the mass consumption from the following three aspects. (i) As mentioned above, since mass has the ability of doing work, it is rather natural that, when an object's mass does *positive* work, the object's mass thereby *decreases* (one can clearly perceive and easily comprehend this aspect through the considerable and explicit help from the comparable concept in classical physics: when a body does *positive* work, the available energy of the body thus *decreases*). (ii) As long as one has known or heard of the famous mass-energy equivalence equation, which is $E = mc^2$, as just mentioned above, he or she can clearly and easily comprehend the mass consumption, because it is inherently connected to the mechanism of this famous equation. To be further explicit, both the mass consumption and this famous equation are attached onto the same thing—the law of mass doing work, because the mass consumption, being caused by *mass doing positive work*, directly and totally comes from this law; because the mechanism of this famous equation, revealed with this law, is the maximum ability or capacity of a rest mass m *doing positive work*, as presented in the section above. That is, both the mass consumption and this famous equation have the same mechanism—*mass doing positive work*. As a result—as a clear and noticeable result in fact, this great and famous equation turns out to be a considerable and explicit help that can enable one to understand the mass consumption easily and quickly. (iii) The fully recognized and completely accepted concept and equation of the relativistic kinetic energy of an object with rest mass m can also provide substantial and explicit help for one to understand the mass consumption. This is because the law of mass doing work directly and totally comes from this concept and equation, as pointed out above; and because the mass consumption is simply the product of an application of this law, as just mentioned above. All in all, with and via the substantial and explicit help from the three aspects above, it seems rather rational or quite realistic to conclude or believe that one will have no difficulty perceiving and understanding the mass consumption, or at least have no difficulty realizing the objective and real existence of the mass consumption.

After seeing the mass consumption introduced above, some careful readers, especially some dear readers who have rich knowledge in physics, may think of relativistic mass (the so-called relativistic mass is an important concept formed within the paradigm of special relativity.

This concept says that: the mass of an object increases with the increase in its velocity, and the mass of an object becomes infinitely large when the object infinitely approaches the speed of light; that is, mass increase with speed. So 'relativistic mass' is often simply said as 'rest mass is least' in the various materials on special relativity); and these readers might have noticed that the mass consumption and relativistic mass are opposite. And so, it seems better that I should provide a relevant clarification here for avoiding possible confusion. This clarification is: there are fundamental and obvious differences between the mass consumption and relativistic mass.

Concisely, these differences are reflected in the following three fundamentally important aspects. (i) The mass consumption is inherently connected with the mechanism of the famous mass-energy equivalence equation, because the theoretical basis of the mass consumption, the law of mass doing work, has also revealed this mechanism, as introduced in the section above. On the contrary, relativistic mass has nothing to do with the mechanism of this famous and great equation, because relativistic mass was the product long before the law of mass doing work had been discovered. Moreover, relativistic mass literally prevents from revealing the mechanism of this famous and great equation. In other words, the mechanism of the famous mass-energy equivalence equation (thus this famous equation) has to say NO to relativistic mass, believe it or not. (ii) The mass consumption is totally consistent with such a fundamental principle in classical physics: when a body does *positive* work, the available energy of the body thus *decreases*. In contrast, relativistic mass is literally at odds with this fundamental principle; that is, according to relativistic mass, when a mass does *positive* work, the mass thus *increases*, which is obviously and flatly ridiculous. (iii) The mass consumption is the indispensable theoretical basis of the new theory that reveals *why* time runs slower at high speed (this theory, which is mechanism-revealed scales relativity theory, has been concisely mentioned in the second section of chapter four); whereas relativistic mass turns out to be an absolutely impassable obstacle to developing such a new theory—that is, within the paradigm of relativistic mass, it is definitely impossible for our human beings to know the secret of *why* time runs slower at high speed. All in all, with and through these fundamental and obvious differences, it seems rather rational that one could get rid of the hindrance or interference from relativistic mass in comprehending the mass consumption.

Chapter 7

After Inspecting All the Three Pillars of the Big Bang Theory

[The Window on This Chapter]

After comprehensive and penetrating inspection of <u>all</u> the three pillars of the big bang theory, the long-standing controversy over this theory is completely over.

After inspecting the crucial pillar of the big bang theory (in chapter one), it turns out that this crucial pillar is not, and cannot be, valid in truth—that is, it turns out that this crucial pillar cannot stand its ground at all; this crucial pillar turns out unable to be tenable, simply because the *prerequisite* for this crucial pillar to be valid **is not, and cannot be, satisfied** at all. (Related question and answer: what is the simple and clear, also feasible and effective, method to know and comprehend that the *prerequisite* for the crucial pillar of the big bang theory to be valid **is not, and cannot be, satisfied** easily and quickly, also impressively and explicitly? Or what is the take home message about that this *prerequisite* **is not, and cannot be, satisfied** at all? Answer: as long as one thinks of or asks the simple and clear question, like: **where does the gravitational redshift in the vast universe go?** This is because this *prerequisite* must require, necessitate and rely on: there is virtually no gravitational redshift at all in the vast universe; that is, all the gravitational redshift in the vast universe simply or virtually evaporates. Or more directly, thus more effectively, as long as one thinks of or perceives the straightforward and inescapable question, like: where does the **great** gravitational redshift caused by the **huge** amount of **dark matter** in the universe go? This is because this *prerequisite* must be totally blind to the **great** gravitational redshift caused by the **huge** amount of **dark matter** in the universe.)

Commentator A: yes, that's definitely correct. Chapter one has indeed provided sufficient, also undeniable or irrefutable, evidence that explicitly shows such a solid conclusion: the *prerequisite* for the crucial pillar of the big bang theory to be valid **is not, and cannot be, satisfied** at all. This solid conclusion has no choice but to reveal or display such an inevitable, also unavoidable, conclusion: the crucial pillar of the big bang theory is not, and cannot be, valid in truth. Moreover, this inevitable and unavoidable conclusion has been further substantiated by the related, more specific and detailed evidence and knowledge introduced in chapters four, five and six.

Commentator B: since it turns out that the crucial pillar of the big bang theory is not, and cannot be, valid, this crucial pillar has actually collapsed, either in essence or in truth or in both. The actual collapse of this crucial pillar virtually or largely indicates that the long-standing controversy over the big bang theory is basically over.

After Inspecting All the Three Pillars of the Big Bang Theory

Commentator C: I totally agree with the comments of the two commentators above. The actual collapse of the crucial pillar of the big bang theory is more or less equivalent to that this theory has essentially collapsed.

After inspecting the necessary pillar of the big bang theory (in chapter two), it turns out that this necessary pillar is not, and cannot be, valid, either in fact or in truth or in both, simply because the *prerequisite* for this necessary pillar to be valid **is not, and cannot be, satisfied** at all. (Related question and answer: what is the simple and clear method or way to know and comprehend that the *prerequisite* for the necessary pillar of the big bang theory to be valid **is not, and cannot be, satisfied** easily and quickly, also impressively and explicitly? Answer: the *prerequisite* for the necessary pillar of the big bang theory to be valid is the hypothesis or conjecture that the assumed big bang was the *only* source of the cosmic microwave radiation measured nowadays. However, as analyzed and unveiled in chapter two, the observed features of gamma ray bursts have explicitly revealed, also unavoidably displayed, such a hard *fact:* in the universe, even merely in the Milky Way galaxy, there are *numerous* celestial bodies that are not only the sources of gamma ray bursts at present, but also the main or important sources of the cosmic microwave radiation measured nowadays. Clearly, even obviously, also undeniably or irrefutably, this hard *fact* explicitly, also unavoidably, shows that the *prerequisite* for the necessary pillar of the big bang theory to be valid **is not, and cannot be, satisfied** at all.)

Commentator A: yes, that's clearly and definitely true. Chapter two has indeed provided sufficient, also undeniable or irrefutable, evidence that explicitly shows such an inevitable conclusion: the *prerequisite* for the necessary pillar of the big bang theory to be valid **is not, and cannot be, satisfied** at all. This inevitable conclusion has no alternative but to reveal or display such a solid, explicit and unavoidable conclusion: the necessary pillar of the big bang theory is not, and cannot be, valid in truth. Moreover, this explicit and unavoidable conclusion has been further substantiated by the related, more specific and detailed evidence and knowledge presented in chapters five and six.

Commentator B: since it turns out that the necessary pillar of the big bang theory is not, and cannot be, valid, this necessary pillar has actually collapsed, either in essence or in truth or in both. The collapse of this necessary pillar, when combined with the collapse of the crucial pillar of the big bang theory just mentioned above, actually indicates that the long-standing controversy over the big bang theory is practically or essentially over.

Commentator C: I entirely agree with the comments of the two commentators above. The collapse of the necessary and crucial pillars of the big bang theory is actually equivalent to that this theory has collapsed in truth.

After inspecting the theoretical pillar of the big bang theory (in chapter three), it turns out that this theoretical pillar is not, and cannot be, valid, thus is not, and cannot be, reliable, simply because the *prerequisite* for this theoretical pillar to be valid thus reliable **is not, and cannot be, satisfied** at all. (Narrator: because the theoretical pillar of the big bang theory is general relativity, the prerequisite for this theoretical pillar to be valid thus reliable is the prerequisite for general relativity to be valid thus reliable. Clearly, even obviously, also undoubtedly and irrefutably, the prerequisite for general relativity to be a valid thus reliable theory is that general relativity must have the ability to solve the most fundamental problem in front of itself, which is *why* space and time are variable thus relative in a gravitational field, or *why* time runs slower and *why* length becomes shorter in such a situation. However, as analyzed and revealed in chapter three, there is sufficient and explicit evidence showing and witnessing that general relativity doesn't and can't have the ability to solve this most fundamental problem at all. Consequently, also unavoidably, the prerequisite for general relativity to be valid thus reliable **is not, and cannot be, satisfied** at all; that is, the *prerequisite* for the theoretical pillar of the big bang theory to be valid thus reliable **is not, and cannot be, satisfied** at all.)

Commentator A: yes, that's definitely true. The careful and penetrating inspection carried out in chapter three has sufficiently and clearly shown or displayed, also unavoidably pointed to such a solid, explicit and inescapable conclusion: the theoretical pillar of the big bang theory is not, and cannot be, valid, thus is not, and cannot be, reliable, simply because the *prerequisite* for this theoretical pillar to be valid thus reliable **is not, and cannot be, satisfied** at all. More than that, this solid, explicit and inescapable conclusion has been further substantiated, corroborated and confirmed by the related, more specific and detailed evidence and knowledge introduced in chapter four.

Commentator B: because it turns out that the theoretical pillar of the big bang theory is not, and cannot be, valid, thus is not, and cannot be, reliable, this theoretical pillar has actually collapsed (or at least is seriously shaky in truth). The actual collapse (or serious shake) of this theoretical pillar, when combined with the collapse of the crucial and necessary pillars of the big bang theory just mentioned above, indicates that the long-standing controversy over the big bang theory is actually over.

Commentator C: I completely agree with the comments of the two commentators above. The collapse of all the three pillars of the big bang theory (its crucial pillar, its necessary pillar and its theoretical pillar) has clearly and unavoidably shown that the big bang theory has collapsed, either in essence or in truth or in both. The collapse of the big bang theory has no choice but to tell us: the long-standing controversy over the big bang theory is over, is finally over, and is completely over.

The final response from the big bang theory: after seeing, pondering and contemplating the solid and explicit, also unavoidable or inescapable, conclusions obtained from inspecting all my three pillars, I totally agree with and completely accept these conclusions. Above all, I entirely agree with all the comments made by all the commentators above, because these comments are rational and objective, pertinent and to the point, clear and concise, highly insightful and penetrating, also quite comprehensive and far-reaching.

GLOSSARY

Acceleration: the rate at which the velocity of an object changes with time.

Astronomy: the branch of science that studies the sun, stars, planets, moon, etc.

Astrophysics: a branch of astronomy that deals with the physical and chemical structure of the stars, planets, etc.

Black hole*: a region of space having a gravitational field so strong that no matter or light can escape. (*Such a definition of black hole is the impassable obstacle to revealing the secrets of dark matter!)

Color index: the ratio of the wavelength of violet to red light.

Compton scattering: the experiment in which the high-energy rays of light, X-rays, were used to strike electrons.

Cosmic microwave background (CMB), or cosmic microwave background radiation (CMBR): the electromagnetic radiation (its wavelength being a few centimeters) that is assumed or hypothesized as a remnant from an early stage of the universe in Big Bang cosmology.

Cosmic rays: also known as cosmic particles, made of electrons, protons, gamma rays and atomic nuclei, are the energetic particles originating outside of the earth. Cosmic rays, except gamma rays (being a form of light with the shortest wavelength) that travel at the speed of light, travel at nearly the speed of light.

Cosmological constant: an important constant originally in general relativity.

Cosmology: the branch of science that studies the universe, such as its origin, structure and development.

Crux: the most important or difficult part of a problem, a matter or an issue.

Dark matter: the huge amount of matter that exists in the universe, such as in galaxies and clusters. The amount of dark matter is estimated to be about 5.5 times of that of the ordinary matter (such as the matter of stars and planets) in the universe—that is, about 85% of the total matter in the universe is dark

GLOSSARY

matter. The existence of dark matter cannot be observed directly but can be detected by its obviously noticeable, enormous gravitational effects.

Electromagnetic radiation: the phenomenon of electrons emitting photons.

Electron: a stable elementary particle with negative electric charge that orbits the nucleus of an atom.

Event horizon: the boundary of a black hole.

Field: the area or space within which a specified force can be felt or has an effect; for example, the earth's gravitational field is the space in which the earth's gravity can be felt or has an effect.

Frequency: the number of waves per second when it is used to describe a wave.

Gamma rays: the highest energy and the shortest wavelength of electromagnetic radiation, which can be generated by nuclear reactions.

Gamma ray bursts (GRBs): the extraordinarily intense bursts or flashes of gamma rays in a very short time from an unknown source (or at least unknown within the paradigm of modern physics). These bursts, coming from different parts of the sky and occurring every day, can last from a fraction of a second to up to a few minutes. The amount of energy released in a (large) gamma ray burst is equivalent to all of the energy stored in the sun, so GRBs are known to be the most powerful explosions in the universe.

General relativity or Einstein's theory of general relativity: the theory developed by Einstein in the early 20th century, it describes and explains the force of gravity with the curvature of a four-dimensional space-time. This theory tells us that space and time are variable thus relative in a gravitational field, and that time runs slower in a gravitational field. General relativity is universally known to be seriously inconsistent with quantum mechanics.

Gravitational redshift: the redshift of light due to a gravitational field.

Gravitational scales of space and time: the space scale and time scale in a gravitational field.

Gravity: the force that attracts objects in space towards each other; the gravity of the earth makes things fall to the ground when they are dropped.

Impulse frequency of an electron: the number of complete impulses of an electron per second.

Impulse period of an electron: the time taken per complete impulse of an electron, which is the inverse of the impulse frequency of the same electron.

GLOSSARY

Mass: the quantity of matter that an object contains; the mass of an object is measured by its acceleration under a given force or by the force exerted on it by a gravitational field.

Mass consumption: a basic principle or direct result that comes from the newly discovered and verified law of mass doing work. This basic principle shows that the mass of an object decreases with the increase in its velocity, by revealing why and how the object's mass is being consumed due to its doing positive work.

Mass-energy equivalence equation: which is $E = mc^2$, where c is the speed of light, m is the rest mass of an object, and E is the rest energy of the object.

Mechanism-revealed black holes[**]**:** the black holes revealed and explained with the newly developed mechanism-revealed black hole theory, which is a fundamentally new black hole theory. According to this new black hole theory, the definition of a mechanism-revealed black hole is: a mechanism-revealed black hole is a hugely massive celestial body that reduces the scales of length and time in its vicinity to such an extent that all visible light becomes invisible—all the visible light entering the region of the black hole becomes invisible; all the light emitted from the black hole is also invisible in the region of the black hole, resulting in this region looking totally black to the observers far away from it. So a (mechanism-revealed) black hole is simply a hugely massive celestial body; the mass of a (mechanism-revealed) black hole is greater or equal to several dozens of times that of the sun. ([**]It turns out that only mechanism-revealed black holes can unveil the secrets of dark matter by revealing its constituents and fundamental nature.)

Mechanism-revealed black hole theory: a newly developed theory of black hole that reveals the mechanism and essence behind the observed phenomenon of black holes; the black holes revealed and explained with this new black hole theory are referred to as mechanism-revealed black holes (for the definition of a mechanism-revealed black hole, please see the above item).

Mechanism-revealed gravitational theory: a newly developed and verified theory of science that solves the fundamental problem of *why* and how space and time are variable thus relative in a gravitational field, by revealing and determining *why* and how their scales are variable thus relative in such a situation. That is, this new gravitational theory unveils *why* time runs slower and *why* length becomes shorter in a gravitational field by revealing the mechanism behind these two *whys*.

GLOSSARY

Mechanism-revealed scales relativity theory: a newly developed and verified theory of science that unveils *why* time runs slower and *why* length becomes shorter in the situation of high speed, and that reveals the mechanism behind the two postulates of special relativity.

Mechanism-revealed theory: a theory of science that finds the mechanism behind its describing phenomena.

Momentum: a quantity of motion of a moving object, measured as its mass multiplied by its velocity.

Momentum conservation (law): the law of science which states that the total momentum of the objects of a system is constant if there are no external forces acting on the system. In such a system, the total momentum of two objects before a collision is equal to the total momentum of the two objects after the collision.

Nucleus: the positively charged central core of an atom, consisting of protons and neutrons, contains nearly all its mass.

Orbiting velocity of an electron: the velocity along the direction of the position of the equilibrium orbit of the electron.

Photon: a tiny and discrete quantum or particle of light (photons: the tiny and discrete quanta or particles of light).

Photoelectric effect: the phenomenon of electron ejection by light; that is, when high-energy light shines on a metal surface, electrons are ejected from the metal surface.

Planck-Einstein equation: which is $E = hf$, where h is Planck constant, f is the frequency of a photon, and E is the energy of the photon.

Planck's quantum theory or quantum hypothesis: the idea that light (or the energy released in electromagnetic radiation from electrons) can be emitted or absorbed only in discrete quanta—photons, whose energy is proportional to their frequency (as shown in the above item).

Prerequisite: an important thing required as a necessary or indispensable condition for something else to happen, exist, or be done.

Proportional: 'Y is proportional to X' means that Y is equal to X being multiplied by any constant number. For example, $Y = 3X$ is an expression of 'Y is proportional to X'.

GLOSSARY

Quantum: a discrete, indivisible quantity or unit of energy, especially the energy released in electromagnetic radiation from electrons.

Quantum mechanics: the theory developed (by Schrodinger and Heisenberg in the 1920s) based on the assumption that all forms of energy out of electrons are released in discrete units called quanta, which are known as photons nowadays. Since quantum mechanics is actually wave mechanics, quantum mechanics is, by using wave mechanics, to describe how electrons radiate photons.

Redshift of light: the measured wavelength of light becomes longer and longer.

Relativistic mass: a concept that comes from special relativity. This concept says that the mass of an object increases with the increase in its velocity, and the mass of an object becomes infinitely large when the object infinitely approaches the speed of light. So this concept is often simply stated as 'mass increase with speed'.

Space-time: the concept of one-dimensional time and three-dimensional space (being intermingled together) is regarded as a four-dimensional system.

Special relativity or Einstein's theory of special relativity: a theory developed by Einstein at the beginning of the 20th century; it has two core concepts, time dilation and length contraction, respectively for interpreting time runs slower and length becomes shorter that appear in the situation of high speed. This theory initiates the era in which time and length are variable thus relative at different speeds.

Singularity: a mathematical point predicted by general relativity, whose size is zero or infinitely close to zero and with infinite density and infinite temperature (at this point, the curvature of space-time described by general relativity becomes infinite. Moreover, general relativity itself breaks down at singularity; that is, general relativity itself is no longer workable or applicable at singularity).

The GZK paradox: one of the long-term unsolved fundamental puzzles in physics or astrophysics. The GZK paradox comes from the GZK limit (computed by Greisen, Zatsepin and Kuzmin in the 1960s), which is a theoretical upper limit on the energy of cosmic rays from a large distance. The GZK limit says that the distant cosmic rays with energies greater than a certain threshold value should never be observed on the earth. However, a number of observations appear to indicate cosmic rays from distant sources with energies above the threshold value. So the key to solving the GZK paradox lies with finding out the sources of these cosmic rays in (our) the Milky Way galaxy.

The law of an orbiting electron with periodic impulses emitting photons: a newly discovered and verified law of science that solves the fundamentally important problem of why there are quantum states, by revealing the quantum mechanism of why and how photons, being the tiny and discrete quanta or particles of light, get their velocity c (the speed of light) from the electron emitting them.

The law of mass doing work: a newly discovered and verified law of science. The core principle of this law is that the amount of energy in the mass of an object is measured by the amount of work done by the object's mass. This law shows that, when the velocity of an object is increased, the object's mass does positive work, the object thus loses the same amount of energy as that of the work done by the mass of the object from and by consuming its mass. As a result, an object's mass doing positive work causes a corresponding decrease in the object's mass.

The mechanism of the famous mass-energy equivalence equation: the rest energy of an object, being the total energy contained in the rest mass of the object, is equal to the maximum ability of the object's mass doing positive work. (This famous equation is $E = mc^2$, where c is the speed of light, m is the rest mass of an object, and E is the rest energy of the object.)

The postulate of 'equivalence principle': this postulate says that gravitational force has the same effect in increasing the velocity of an object as other traditional forces (this postulate, being a fundamentally indispensable postulate of general relativity as its fundamental and crucial foundation, is known to be the heart and soul of general relativity. Moreover, this postulate is widely known to be one of the most important, most influential and most famous postulates in modern physics).

Ultrahigh-energy cosmic rays: the cosmic rays that have extraordinarily high energy. The mysterious source(s) of ultrahigh-energy cosmic rays has been widely and fully recognized by the related scientific communities as one of the most fundamental mysteries in physics or astrophysics for quite a long time.

Wavelength: the distance between two adjacent crests or two adjacent troughs.

INDEX

Acceleration 104, 108
All the three pillars of the big bang theory 127, 131
   See also the three pillars of the big bang theory
Aristotle 93
Assumptions, hypotheses and postulates (AHPs, for short) 47, 59
Astronomy 5-6, 16, 21, 25, 49, 73, 75, 82, 95, 98, 106
Astrophysics 16, 20-21, 24-25, 75, 106
Average density 75, 104
   See also Density

Big bang singularity (The) 38-40

Celestial bodies 13-14, 16-18, 21, 23, 25-26, 39, 65, 71, 74-76, 79, 96, 100, 102-04
   See also Heavenly bodies
Celestial body 46, 50, 53-54, 56, 58, 60-61, 63, 65, 69-73, 75-76, 87, 95, 100-02, 104, 109
Classical physics 116-17, 122-23, 125
Color index 72
Compton, Arthur 93
Compton scattering 91, 93
Continent of North America 108, 119
Contour line of space and time 52, 60, 85
Contour lines of space and time 50, 52, 60, 83
Cosmic microwave background 13, 17-18, 23, 26

Cosmic microwave background (radiation) 13, 17, 23, 26
Cosmic microwave background radiation (CMBR) 13, 18
Cosmic microwave radiation 13-19, 23-24, 26, 39-40, 129
Cosmology 5-6, 49, 95, 98
Crucial pillar of the big bang theory, The 1-3, 10, 127-30
Crux 21, 97, 105
Current method of calculating the momentum of photons 77, 88-93
   See also Momentum; Momentum conservation law

Dark energy 49, 97
Dark matter 1, 5-6, 7-8, 11, 41-42, 44, 49, 65, 76, 86, 95-106, 109-14, 128
Density 22, 36, 38, 75-78, 103-04, 109
Dimensions 53
Doppler effect 2-3, 7, 9-10

Earth 1-5, 10-11, 20, 22, 24, 33, 45, 50, 52, 56, 59-60, 62-63
Einstein 27, 89, 93, 121
Einstein, Albert 27, 121
Einstein's theory of general relativity 98
   See also general relativity, Postulate-based general relativity, & the theory of general relativity
Electron or electrons 20, 24, 89-92
Electron-photon momentum relationship 90-91
Elements 104

INDEX

Encyclopedia of physics, The 21, 25, 97
Equilateral triangle 85
Event 74, 76-77, 86-88, 92-94, 109, 119, 123
Event horizon 74, 76-77, 86-88, 92-94, 109
Event horizon of a black hole 77, 86-87, 109
Expanding universe 1-3, 5, 7, 9-10
Experimental tests 33, 121

First postulate of special relativity, The 44
Force 47-48, 53, 87, 104, 107-08, 117
Forces 47, 107-08
Franco, Fernando 48
Frequency 22, 55-56, 58, 68, 81, 89-90, 105

Galaxies 1-2, 5-11, 14, 16, 82, 96
Galaxy 5-6, 8, 13, 17-18, 23-26, 39, 75, 82, 96, 102, 104, 129
Galilei, Galileo 93
Gamma ray bursts 13, 16-17, 20-25, 39, 76-77, 86, 109, 129
Gamma rays 18, 20-22, 24-25, 76
General knowledge or common sense 5, 33, 43
General relativity 27-37, 40, 42, 45-49, 58-59, 63-64, 66-67, 73-78, 86-87, 94, 98-99, 106-09, 112-13, 130-31
  See also Einstein's theory of general relativity, Postulate-based general relativity & the theory of general relativity
George 69
Gravitational field 2, 4-6, 8, 11, 27-28, 30-35, 40-42, 45-56, 58, 60-64, 66-67, 69, 71, 73, 78, 81, 83, 85, 96, 98-99, 101-02, 106-07, 111-13, 130

Gravitational fields 2, 5, 7, 9-11, 48, 96
Gravitational force 47, 53, 107-08
  See also Force
Gravitational lens 6, 96, 103, 111
Gravitational light bending 60-63, 82-85, 103, 110-11
  See also light bending
Gravitational redshift 1-11, 42, 55-59, 78-82, 103, 110, 128
  See also Redshift
Gravitational scale contour line of space and time 52, 60, 85
Gravitational scale contour lines of space and time 50, 52, 60, 83
Gravitational scales of space and time 50, 73
Gravity 6, 46, 86-88, 96, 103, 111
GZK paradox 25, 76, 86, 109

Harvard tower experiment 4, 45, 59
Hawking, Stephen 66, 74, 86
Heavenly bodies 96

Impulse period 91
Indispensable 19, 27-28, 31-33, 38, 41-43, 48-49, 58, 68, 86, 107, 114, 125
Indispensably 31, 58, 94
Infinite density 36, 38, 77-78
  See also Density
Interference 126
Invisible light 68-69, 74
Israel, Werner 67

Kerr, Roy 67

Length becomes shorter 27-28, 30, 32-35, 40-47, 49, 51, 54-55, 58, 63, 66-67, 78, 98-99, 106-07, 112-13, 130
Light bending 6, 33, 45, 60-63, 82-85, 103, 110-11

INDEX

Light bending around the sun 33, 45

Mass consumption 53, 107-08, 123-26

Mass-energy equivalence equation 104, 114-15, 118, 120-25
See also The mechanism of/behind the (famous) mass-energy equivalence equation

Mass increase with speed 125
See also Relativistic mass

Measurement 55, 59

Mechanism-revealed 41, 43, 47-49, 51, 56-68, 70-85, 95, 99-104, 106, 109-12, 125

Mechanism-revealed black hole 65, 71-74, 76-79, 81-83, 85, 100-02, 109-11

Mechanism-revealed black holes 65-66, 71, 73-74, 77-79, 81-85, 95, 100-01, 103, 106, 109-12

Mechanism-revealed black hole theory 65-68, 70-74, 76, 79, 100-01, 103, 106, 109, 112

Mechanism-revealed gravitational theory 41, 47-49, 51, 56-58, 60-62, 65-66, 68, 73, 78, 85, 99, 112

Mechanism-revealed scales relativity theory 43, 47, 126

Mercury's orbit 33, 45

Milky Way galaxy 5-6, 8, 13, 17-18, 23-25, 39, 75, 82, 104, 129
See also Galaxy

Modern physics 5-6, 21, 25, 31, 117

Momentum 48, 77, 87-93, 117

Momentum conservation law 90

Necessary pillar of the big bang theory, The 13-15, 18, 129-30

Newtonian gravitation theory 87

Newton's second law 48, 87, 104

Observational tests 33-34

Oppenheimer, Robert 67

Photoelectric effect 89-90, 93

Photons 20, 48, 60, 62-63, 76-77, 87-94, 109

Planck 89, 105

Planck constant 89, 105

Planck-Einstein equation 105

Postulate-based 47-49, 59-60, 63-64, 66-67, 74-78, 86, 92-94, 106-10, 112-13

Postulate-based black hole 74, 109

Postulate-based black holes 66-67, 75, 77, 86, 106

Postulate-based black hole theories 66, 74-76, 78, 86, 106, 109-10, 112-13

Postulate-based general relativity 48-49, 66-67, 74-78, 86, 106-09, 112
See also Einstein's theory of general relativity, general relativity & the theory of general relativity

Postulate of 'equivalence principle', The 47, 107-08

Postulate of 'invariant scales of length and time', The or Indispensable 31-33, 47, 58, 107

Prerequisite 1-5, 7-10, 13-15, 18-19, 27-30, 34, 42, 49, 76, 100, 105, 112-13, 127-31

Princeton University 4, 59

Proportional 48, 73, 87

Radius of a mechanism-revealed black hole 72, 74, 81, 85

Ratio of visible light boundary wavelength 72

Redshift 1-11, 42, 55-59, 78-82, 103, 110, 128

Relationship 43, 62, 79, 90-91

Relationships 14, 30

Relativistic mass 124-26

INDEX

Reveal (& revealed, -revealed revealing, reveals) 10, 13, 16-17, 32, 34-35, 39-51, 54-68, 70-85, 91, 95, 97, 99-112, 114-15, 118-25, 128-30
Root cause 64, 108
Rotation of the long axis of Mercury's orbit 33, 45

Schwarzschild, Karl 67
Schwarzschild radius 73, 85
Science Magazine 24-25, 97
Second postulate of special relativity, The 44
Secrets & secret 51, 120, 126
Self-contradictory 36-38, 40, 77-78, 93-94
Self-evident 28, 43, 108
Singularity 35-40, 77-78
Size of a black hole 73
Space-time 48
Special relativity 44, 94, 124-25
Speed of light 24, 36, 38, 44, 72, 78, 87-89, 94, 104, 114, 118, 125
Star 2-4, 10, 61
Stars 1-2, 4-8, 10-11, 16, 74-75, 81-82, 96, 102-06, 111
Sun 4-6, 8, 11, 22, 24, 33, 45, 50, 56, 59-60, 62-63, 75, 81-82, 104
Supermassive black hole 8, 82

Temperature 36, 38, 77-78
The angle of science 118
The big bang theory 1-3, 10, 13-15, 18-19, 27-30, 32, 34-35, 38, 127-32

The law of an orbiting electron with periodic impulses emitting photons 90
The law of mass doing work 108-09, 114-19, 121-25
The mechanism of/behind the (famous) mass-energy equivalence equation 114, 118-25
The mechanism of/behind the postulate of 'equivalence principle' 107-08
Theoretical pillar of the big bang theory, The 27-28, 30, 34-35, 130-31
Theory of general relativity, The 27, 67
Three pillars of the big bang theory, The 127, 131
Threshold radius 72
Time runs slower 27-28, 30-35, 40-47, 49, 51, 54-55, 58, 63, 66-67, 78, 98-99, 106-07, 112-13, 125-26, 130

Ultrahigh-energy cosmic rays 16-18, 20, 24-26, 76, 86, 109

Vacca, John 21, 97
Velocity 47, 87-88, 94, 107-08, 116-17, 123, 125
Visible light 65, 68-72, 74, 76, 95, 100-02, 109

Wavelength 20, 24, 55-56, 58, 68-69, 72, 81, 90
World's 20 Greatest Unsolved Problems (a book), The 21, 97

ABOUT THE AUTHOR

Bingcheng Zhao, who was born in 1963 in Shandong Province of China, obtained his Ph.D. in 2001 from Washington State University. He is the author of the popular science books: *Why It's Difficult to Understand "A Brief History of Time"*, published in 2016, and *The Long Night Is Over: The Mystery of Dark Matter Has Been Unveiled; The Constituents of Dark Matter Have Been Revealed*, published in 2017. He is also the author of the academic book: *From Postulate-Based Modern Physics to Mechanism-Revealed Physics*, published in 2009; the newly developed and verified mechanism-revealed physics is the key to solving the fundamentally important problems of dark matter and dark energy; and the birth of mechanism-revealed physics actually heralds that the spring of science is coming again.

www.ingramcontent.com/pod-product-compliance
Lightning Source LLC
Chambersburg PA
CBHW031417210526
45464CB00005B/1929